CAPITAL FORMATION IN AGRICULTURE SECTOR OF MANIPUR

CAPITAL FORMATION IN AGRICULTURE SECTOR OF MANIPUR

(1972-1997)

N. Bhupendro Singh

AKANSHA PUBLISHING HOUSE

NEW DELHI - 110 002 (INDIA)

AKANSHA PUBLISHING HOUSE
4649/21-B, Ansari Road
Daryaganj, New Delhi - 110 002
Phones: 23263193, 9811582579
E-mail: ektabooks@yahoo.com

Capital Formation in Agriculture Sector of Manipur

First Published, 2007

ISBN: 978-81-8370-100-6

PRINTED IN INDIA

Published by M.P. Misra for Akansha Publishing House, New Delhi - 110 002 and Printed at Capital Offset Press, Delhi

Dedicated

To my mother

Mrs. N. Chandramukhi Devi

Contents

Foreword

The developments in agricultural sector in Manipur can be summed up as follows.

- Manipur degenerated from a surplus rice producing state to a deficit state.
- Agriculture in Manipur has not been accorded its due importance in terms of investment
- Agricultural output grew at a constant rate indicating how green revolution has by passed it.

Besides growing unemployment, poverty and inequality have posed critical questions about the appropriateness of the paradigm of development policy in the state. It is increasingly realized that contrary to current policy stance, agriculture will continue to play a key role in the economic development of Manipur. However a strategy based on extensive agriculture is no longer sustainable as they is heavy demand on arable land due to urbanization and growth of population. Besides only a small portion of Manipur is suitable for settled cultivation due to topography.

Yet agriculture needs to be made as vibrant and dynamic as possible. The important question is—what role can capital play in a situation characterized by scarce land and abundant labour? The related questions are its trend growth, determinants and effects on agriculture. It is in this context that the interface between capital formation and agriculture productivity calls for more intensive probe. This makes Dr. Bhupendro's work a valuable contribution on capital formation in agriculture in particular and agricultural development in general.

Dr. Bhupendro has examined inter alia the trend of agricultural capital formation both public and private, determinants of private capital formation and their impact on agricultural production—all in the context of Manipur. Besides its time lines and relevance, it helps in bridging a gap in data base. Seldom do we find any substantive research work on such an important aspect of such a small state like Manipur. The construction of a time series of capital formation in agriculture is indeed a welcome addition to our data base. The study of determinants of private capital formation and impact of capital formation on agricultural production is highly analytical with rigour and valuable policy in sights can be gleaned from this finding.

Public capital formation is treated as an exogenous process. Private capital formation is sought to be explained in terms of public capital formation lagged by one period, terms of trade, technology and interaction between technology and terms of trade. A key finding is the relative strength of private capital formation over public capital formation in explaining the variation in agricultural output. At the micro level also 92 percent of farm output variation is explained by land and accumulated capital stock of farmers.

Dr. Bhupendro's work indeed is a pioneering contribution and I am sure that younger research scholars will be encouraged to pursue this aspect with proper perspective. It is a watershed in the study of agricultural production function in Manipur. With this work, the issue has been highlighted than ever, delineating the contours for further work. The work, I hope, will be carried forward to greater heights.

—E. Bijoykumar Singh

Preface

Capital formation has been considered an important sinew in the growth and development of agricultural sector of an economy. The task of developing this sector cannot be accomplished without the judicious combination of factors and efficient utilisation of capital formation. Any shortfall in capital formation, when the demand for food and other agriculture related products has been increasing tremendously due to inexorable rise in population, ought to become a source of concern. This has inspired a number of researchers to conduct an in-depth empirical examination on the quantum of capital formation along with its various aspects beginning with conceptual, definitional, problems associated with measurement and its role in the growth and development of this sector. Such analysis would enable us to understand the past structure and different aspects associated with it. It would also help in projecting the future trend, which would enable to meet the growing demand for food and for achieving overall sustainable rate of growth and development of this sector in particular and the economy in general.

In India the task of estimating agricultural capital formation has been entrusted with Central Statistical Organization (CSO). The CSO estimates published in National Account Statistics (NAS) series, however, provide the aggregated series at national level and for some major states of India. It completely omits the scene prevailing in small states like Manipur. In realising the role played by capital formation in development planning of this sector and complete absence of its quantitative estimates in the CSO series, an attempt has been made to quantify and study various aspects associated with capital formation in agricultural sector of Manipur.

It may be pointed out that, like CSO estimates, the present study shall also be confined to different items of physical durable or tangible capital which are included in the CSO concept of capital formation. This will exclude human and other intangible capital included in broader definition of agricultural capital formation.

The estimated series of agricultural capital formation of Manipur based on CSO coverage and methodology has been used to examine some of the basic issues of research like structure and nature of relationship between public and private capital formation, determinants of private capital formation and impact of public and private capital formation on agricultural production and growth in the state using standard quantitative techniques like correlation and regression analysis.

The estimation of agricultural capital formation in Manipur is, however, subject to many limitations apart from other inherent limitations of CSO estimates - mainly due to the paucity of data. The present study may, however, be regarded as only the first approximation. Further corrections and revisions in the future would be highly imperative with the availability of reliable relevant additional data.

The schemes followed in the present study are as follows: Chapter I deal with the conceptual and definitional aspects of agricultural capital formation and the role of agriculture in economic development of an economy in general. In Chapter II an attempt has been made to outline a brief sketch on agriculture sector of Manipur with special reference to agrarian structure during the period 1972-73 to 1997-98. Chapter III deals with the coverage, methodology and sources of data. In Chapter IV we analyse the structure and quantum of agricultural capital formation in the state and also the trend growth rate during the period 1972-73 to 1997-98. It also deals with the quantum and composition of accumulated capital stock in cultivator households of the four valley districts based on field survey data collected in the year 2000. In Chapter V an attempt has been made to ascertain the factors determining the private capital formation and the nature of relationship between public and private capital formation. In Chapter VI we analyse the impact of capital formation on agricultural production and growth in the state. It also deals with the impact of accumulated capital stock on agricultural production based on field survey data. In Chapter VII a brief sketch of some useful inferences arrived at different stages of analysis and remedial measures to counter various limitations encountered in due course of the study if and when relevant data are available have been highlighted.

The present book is substantially based on the findings of my doctoral research in Economics carried out at the Manipur University, Imphal.

I acknowledge my deep sense of gratitude to Prof. E. Bijoykumar Singh, Head, Department of Economics, Manipur University for his valued advice and suggestion at every stage of analysis. His comments and criticism had helped me immensely in completing the present book. I shall also remain ever indebted to Prof. E. Bijoykumar Singh for the foreword to the present book.

I am also truly grateful to all the teachers and staff of Economics Department, Manipur University for the award of departmental fellowship during my research period. Without their constant encouragement, motivation and appreciative understanding it would not have been possible to complete the research work.

I acknowledge my indebtedness to ICSSR, New Delhi for the award of much needed Partial Assistance fellowship at the time of completion of my research work. The fellowship had helped me immensely in easing the financial pressure at the closing stage of the research which most of the research student encountered.

I express my heartfelt gratitude to all the learned persons, colleagues and friends who helped me directly or indirectly all the way. My special thanks are also due to them.

Last but not the least, I shall remain ever indebted to my mother and my wife, brothers, sisters and sisters-in-law who have swallowed the pain to support me throughout. Their ungrudging support, motivation and encouragement have always been a source of inspiration in my endeavour to complete this book.

—N. Bhupendro Singh

List of Tables & Figures

TABLES

FIGURES

1

Introduction

When man established himself a settled life with agriculture as the main occupation, capital has been used as one of the most important factor of production. Since then the intensity and composition of capital have altered immensely with the advancement of human civilization leading to a continuous changed in the process and techniques employed in the production of goods and services. In every production process more of labour factor is being substituted by capital factor. The productivity of land and labour is being increasingly determined by the quality and quantity of capital used in every production process. In this modern era of civilization the productive capacity or productive base of every economic activity has become more capital intensive rather than its labour size or land resources. The extent of efficacy of all the productive effort ultimately depends upon the size of capital being employed. Thus, it has been widely accepted that the task of developing an economy cannot be accomplished without the judicious combination of factors and efficient utilization of available capital.

Although, theoretically, it may be possible to produce a commodity only with the combination of land and labour but, practically, it is not possible to produce anything without some amount of capital. This is because, inspite of the significant role played in the production process, capital by itself does not stimulate any independent impulse. It only transmits an impulse given by the original productive powers of land and labour.[1] Therefore, one may conclude that, capital is necessary factor but not the sufficient one in the process of production.

Capital is different from all the other natural resources that are the gifts of nature. It is an intermediate product of land and labour. Its own origins, its existence, its subsequent action, are nothing but stages in the continuous working of the true elements—land and labour.[2] Capital is only a tool of production. It abridges the original factors in the production process. That is, capital acts as the medium through which the two original productive powers i.e. land and labour exert their instrumentality.[3] In short, we can say that all forms of capital which acts as a catalyzing agent in the development of an economy are manmade with the application of available natural resources and, therefore, its stock can be increased by human effort only.

There is no precise definition of capital. It has been defined in a number of ways depending upon the nature and scope of the concept in the process of production. "A unique definition of capital for understanding the nature of economic process is not possible since the nature of the economic process itself is not unique."[4] To those economists who are concerned mainly with the problem of static equilibrium, the question of defining capital does not present a big problem.[5] In such economy capital can be accurately measured and we can know precisely what is meant by 'keeping capital intact'.[6] In a stationary economy the form, function and coverage of capital in the process of production are quite limited. But "so long as the economy is undergoing any sort of change, and capital goods are altering their form, we cannot know exactly to what extent capital is increasing or decreasing."[7] In today's dynamic economy the problem of defining capital has become infinitely more difficult because of its varied forms, functions and wider range of coverage. They keep on changing as a consequence of technological advancement. Perhaps in the prevailing pace of change of economic system, we should abort the quest for precise definition of capital and its true measure. Instead it would be more appropriate to concentrate on defining a more reliable pointer applicable to particular production process.

Capital has been defined in a number of ways. Bohm-Bawerk (1948) defined capital as a "produced means of further production."[8] A United Nations Study (1955) defined as "those goods resulting from economic activity, which are used for future production of other goods."[9] Colin Clark (1957) defined as "those reproducible wealth which are used for the purposes of production."[10] Thus, by capital we meant all those assets or reproducible wealth or goods, which are used directly or indirectly for augmenting production over a period of time in the future.

Capital as a produced means of further production takes different forms based on the nature of economic function it performs. They take the form of tangible assets or physical or material capital such as plants, raw materials, machinery and equipments, all kinds of construction, etc. and intangible or human capital which are embodied in oneself such as education, technical training and managerial skill and knowledge. In short, capital includes both physical as well as human capitals.

In the earlier part of our civilization, of the two components, physical capital used to play a dominant role in the functioning of economic activities. But in the recent years human capital has been considered as one of the most important components of capital. "With time stretched out to the farthest ends, the investment in human being has become an important form of capital, and the investment in human being may not necessarily be restricted to the improvement of metal equipment and physical dexterity, it may as well include the crude physical capacity of the human being too, since it too is yoked to production."[11] All the new forms of material capital innovated or designed to suit the changing environment in the production system are coming only from the cranium of those educated skill labourers. Furthermore, a trained labour is a better ally to physical capital and performed better than untrained labour thereby resulting into increase in production. The education and training part acquired by labour force has also been considered as capital.[12] Human capital not only determines the quality and composition of physical capital but also raises the efficiency and quality of labourers and hence productivity. Though the importance of human capital is gaining momentum no appropriate yardstick has so far been formulated to measure it.

It is well-known fact that capital—physical as well as human capital - plays an important role in the process of production in particular and development of an economy in general. Now the pertinent question is how do we enlarge the stocks of capital? The size of the capital could be enlarged if and when there is a net addition to the existing stock of capital. It could be achieved only when the rate of capital formation is accelerated consistently which in turn depend upon the extent of investment and saving in the economy.

1.1 Capital Formation

Capital formation in any productive system means net addition made to the existing stock of capital goods in a given period of time. It not only includes all kinds of material capital like machine, tools, plants, instruments, constructions, etc. but also non-material goods like knowledge, technical skill and health of the people. The net addition in the existing stock of capital could be arrived when there is significant rate of saving and investment concurrently. "The meaning of 'capital formation' is that society does not apply the whole of its current productive activity to meet the needs and desires of immediate consumption but directs a part of it to the making of capital goods; tools and instruments, machines and transport facilities, plant and equipment—all the various forms of real capital that can significantly increase the efficacy of productive effort. The term is sometimes used to cover human as well as material capital; it can be made to include investment in skill, education and health a very important form of investment."[13]

The investment in physical asset, which leads to capital formation, is well known and need not be reiterated further. Investment in human beings means knowledge and skill formation. The level of education that a person acquired would reveal the extent of knowledge explored and accumulated by him. In all the sectors of the economy, an educated trained labour is more effective in the process of production than an uneducated and untrained labourer. The skill to handle more sophisticated machinery and equipments that one acquires from training and education is a part of capital formation. It implies that, apart from physical capital formation, human capital formation has also become one of the most vital components of the capital formation.

Capital formation in an economy is mainly the result of abstinence of a part of present consumption to be utilized in future production. The net addition made to the existing stock of capital is possible only when a part of the current income is transferred for the formation of capital which can enlarge future production. The essence of capital formation is the diversion of a part of current income by sacrificing present consumption and investment in a more productive way in the future.[14] It may be noted that the types of investments qualified to be included in the capital formation are only those investments incurred in the formation of real physical capital or human capital. While investments incurred on acquisition of financial assets are to be excluded, as they do not enlarge productive capacity directly.[15] In other words, capital formation is the outcome of those investments incurred on the addition or creation of physical assets or human capital which are directly involved in the process of production.

There are various steps involved in the process of capital formation. Meier (1970) broadly classifies the steps involved in the process of capital formation into the following three steps;

(i) an increase in the volume of real saving so that resources can be released for investment purposes;

(ii) the channelising of savings through a finance and credit mechanism, so that investible funds can be collected from a wide range of different sources and claimed by investors;

(iii) the act of investment itself, by which resources are used for increasing the capital stock.[16]

The saving a part of current income and channelising it to the creation of physical assets or goods, which are more productive, is the essence of capital formation. This implies that, "capital formation requires an act of investment as well as a capacity to save."[17] In order to accelerate the rate of capital formation the rate of investment should be increased which is directly related to the amount of saving accumulated in the economy. These show that saving is directly proportional to capital formations only if and when savings are converted into real investment. Even though there were a high rate of saving, if they are not utilized in the creation of physical assets to augment future production the rate of capital formation would continue to be low.

One can easily grasp this scenario in underdeveloped countries. These countries are manifested with vicious circle of low income, low saving, low investment and low rate of capital formation, which ultimately leads to low income. Over and above the meagre amount of saving associated has not been utilized properly in the creation of capital goods, which would be used in further production of goods and services. These countries lack strong political will on the part of government to induce the general public for more saving and lacks proper infrastructure needed for channelising the saving so generated to the desired areas. As a result the rate of capital formation in these countries continues to be low and hence inordinate delay in development process.

Furthermore, the rate of human capital formation in these countries is also very low over and above the low rate of material capital formation. In other words, the rate of economic growth in the underdeveloped countries is quite low no matter how much physical or material capital is applied in the process of production. This is mainly because of low rate of human capital formation. As Meier (1970) has said "in many newly developing countries the absorptive capacity for physical capital has proved to be low because the extension of human capabilities has failed to keep pace with the accumulation of physical capital".[18] Both material and human capital formation should grow in a pace ensuring optimum mixed between them. Investment in human capital leads to changes in the composition of physical capital, qualitatively and quantitatively, in the economy which ultimately influence the pace of development of that country. In this modern era of civilization investment in human capital formation is one of the key factors in determining the pace of development of a country.

1.2 Capital Formation in Agriculture

Since the days of Adam Smith, almost all economists regarded capital formation as one of the most important determinant of economic growth and development. Nurkse opined that, the vicious circle of poverty in underdeveloped countries could be broken through capital formation. The famous Harrod-Domar equation, which was incorporated into the planning models of many underdeveloped countries including India, stated that the growth rate was dependent upon the rate of investment and the capital-output ratio. For a given capital-output ratio, the higher the rate of investment higher is the rate of growth. Similarly, Arthur Lewis asserted the importance of capital formation in economic development in his 'Economic Development with Unlimited Supplies of Labor'. According to him "the central problem in the theory of economic development is to understand the process by which a community which was previously saving and investing 4 or 5 percent of its national income or less, converts itself into an economy where voluntary saving is running at about 12 or 15 percent of national income or more. This is the central problem because the central fact of economic development is rapid capital accumulation (including knowledge and skills with capital)."[19]

The growth and development of an enterprise in particular and economy in general is determined ostensibly by the level of technology and efficiency of market mechanism. While the level of technology determines the quantum of output that can be produced from a given input resources, the size of the market is determined by the volume of production and productivity.[20] This implies that "capacity to buy means capacity to produce."[21] However, an improvement in the level of technology and efficiency of market mechanism is grossly inclined on both quality as well as quantity of capital employed in the economy. It means growth and productivity of an enterprise is mainly dependent upon the degree to which capital is employed in production processes.[22] When the stock of capital increases, naturally the technical form of it also changes as a consequence of technological progress. Similarly, it would also help to remove market imperfections by creating economic and social overhead capital thereby induces the expansion of market.

The government of India realized the importance of capital formation since the inception of economic planning. By adopting the Harrod-Domar growth model in the First Five Year Plan, the government of India stated "the key to higher productivity and expanding the level of income and employment lies really in stepping up the stock of capital at its disposal, i.e. on the amount of land per capita and of production equipments, tools, and implements, factories locomotives, engines, irrigation facilities, power installations and communication."[23] It was reiterated in the Second Five Year Plan as well. In short, a significant amount of capital formation is required for achieving the much-desired goal of growth and development of every economic system. Agriculture sector is not an exception. Agriculture being the primary sector of every underdeveloped economies, it is highly imperative to exert significant thrust to increase the stock of capital in this sector as well.

Agriculture is the oldest occupation of mankind. It occupies a very important place in the economy of the underdeveloped countries or states. It has been the single largest contributor both in terms of the share of gross domestic product and employment in every developing country's economy including India. This sector not only feeds the entire population but also provides the necessary raw materials required in the development of other secondary sector as well. Gandhi (1997) has identified seven major contributions of agriculture to economic development of these countries. They are, i. Product contribution; ii. Income contribution; iii. Employment contribution; iv. Factor contribution; v. Capital contribution; vi. Trade contribution; vii. Demand contribution.[24] In an underdeveloped economy "if development is to take place and become self-sustaining it will have to start in the rural areas in general and the agricultural sector in particular."[25] In these countries the prosperity of the people is very much interlinked with the development of agriculture sector.[26] "The core problems of widespread poverty, growing inequality, rapid population growth and rising unemployment have their origins in the stagnation and often retrogression of economic life in rural areas."[27] It, therefore, should be given the highest esteem and importance in the growth and development of this sector in every plan formulation. Any sort of negligence would have an adverse impact on production and productivity of this sector in particular and economy in general.

With the growth of the population in geometric progression, the demand for food and other agricultural products has increased considerably. On the other hand, the land-man ratio has been increasingly adverse, as the scope for further extension of arable land seems to have reached the point of saturation. Any attempt to expand the net sown area could be at the expense of ecological harmony, which is against the concept of sustainable development. Indeed ever-burgeoning population exerts tremendous pressure on the size of cultivable land while the fertility of land varies little overtime. Furthermore, the institutional setup[28] that has been inherited from our ancestors also acted as an equally important stumbling block in the growth of agricultural output. In a country/ state where agricultural sector is dominated by marginal and smallholdings the goal of rising agricultural production is unlikely to be realized unless suitable technologies compatible with the prevailing institutional framework have been employed. This is because "an agrarian structure with unequal land distribution is understood to lead to under-utilisation of land and, therefore, low productivity."[29] The employment of suitable technology will enable to achieve optimum utilization of other factor inputs such as land, labour, etc.

In most of the underdeveloped countries this sector has been found to be associated with, in the language of Schultz, traditional agriculture. According to him traditional agriculture means "the state of art and the state of preference and motives for holding and acquiring agricultural factors as sources of income streams have remained approximately constant for a long period—long enough for suppliers and demanders of agricultural factors to have arrived years ago at a particular long run equilibrium—belongs to this class. Its critical economic characteristic is the high price of the sources of permanent income streams from agricultural production."[30] The major attributes of traditional agriculture are the used of labour-intensive technique of production with unchanged primitive technology, low level of skills formation and technical know-how of the labour force. In this type of agriculture, production is mostly for self-consumption and therefore, the extent of marketisation is very limited. A fairly large proportion of the labour and materials used in production and investment is owned by the families and consequently, the degree of monetisation is also very low.[31] In traditional agriculture land-labour-bullock activity is the way of life. The net capital formations is very low and are mostly of traditional type as the rate of investment is negligible. The low rate of investment is mainly due to low saving and backwardness of technology in used. In such cases it is very difficult to meet the growing demand for agricultural products. To achieve the goal of raising agricultural production it has become highly pertinent to transform this tradition-ridden sector into a dynamic modern sector with the application of modern science and technology, new skill and knowledge, and development of more efficient market structure. This, however, could be achieved only when there is concurrently high rate of saving and investment leading to higher rate of capital formation. Any decline in the rate of investment for capital formation would hamper the progress of transformation of agriculture from a predominantly subsistence sector to a surplus generating business sector.[32]

In modern agriculture the principle source of high productivity is capital formation. The components in land and in man that are fixed by nature are generally of secondary importance.[33] The transformation of traditional agriculture into a modern sector required to developed and supply a more profitable set of factors. Development and supply of such factors and learning how to use them efficiently constitute investment in both human and material capital.[34] This increasingly calls for larger and more efficient utilization of capital and its formation in the production process. Increase in capital formation would also lead to increase in the efficiency of labour. It would enable the farmers to take full advantage of the available technology thereby reducing the gulf of available technology and applied technology. This reveals that inputs and technologies used in modern agriculture are highly capital intensive as compared to that of traditional agriculture.

Shifting priority away from agriculture to manufacturing sector before it is fully developed would have an adverse impact on the income and welfare of the people associated with in particular and overall development of the economy in general. A major reason for the relatively poor performance of agriculture in underdeveloped countries has been the neglect of this sector in the development priorities of the respective governments. The neglect of this sector and accompanying bias in favour of investment in the urban industrial economy in turn can be traced largely to the misplaced emphasis on rapid industrialization via import substitution and exchange rate overvaluation that permeated development thinking and strategy during the post-war decade.[35] A questionable notion indoctrinated in our Second Five Year Plan. Shifting of investment toward non-agriculture sector may result in a faster growth in total gross domestic product in the long run. Yet, the growth across sectors is likely to be highly uneven, with non-agricultural sector showing a higher growth rate than agriculture.

Slowing down of agricultural growth would lead to growing income inequality between rural areas and urban areas. A sound development of this sector—foundation of every developing economy—is the crux of sustainable development of an economy. This sector can be a powerful engine for growth if suitable forms of investment for large-scale production are applied. Incentives for guiding and rewarding the farmers are critical. Once there are investment opportunities and efficient incentives, farmers will turn sand into gold.[36] All this implies that capital formation is the pre-requisite for the growth and development of this sector. It sets the pace and pattern of agricultural production in particular and economy in general.

Until recently capital formation in agriculture has been defined in a very crude form. Capital formation in agriculture comprises net addition made to the existing stock of assets like livestock, land reclamation, bunding and other land improvements, irrigation, constructions, orchards and plantation, tools and equipment, etc. Schultz (1964) calls these forms of capital traditional capital. These traditional capital are formed mainly with the direct application of unpaid family labour which can be created without much financial saving.[37] In those productive systems where traditional forms of capital predominate the goal of raising production is unlikely to be realized no matter how fertile the land is.

Schultz defined capital formation in agriculture on a different perspective. According to him, the goal of raising agricultural production would not be feasible only with the traditional forms of capital. Production in abundance of farm products requires that the farmer has access to and has the skill and knowledge to use what science knows about soils, plants, animals and machines.[38] In modern agriculture production activity has been evolved around technology-seed-fertiliser-water combination where human agent plays a significant role. The improvement in the quality of factor inputs including the human agent lies on the level of skill and knowledge acquired in oneself. The production activities in modern agriculture have inclined towards human resource rather than natural resource. Similar to other sector of the economy, human capital formation has also been reckoned in the definition of capital formation in agriculture. It plays an important role in injecting the technological knowledge so as to transform traditional capital into modern capital that is more ally to the vastly diversified and heterogeneous present day economic system.

Investment in agriculture is a very complex process considering the degree of risk and uncertainty associated with this sector. The rate of return on investment in this sector is highly volatile. As a result investors are always reluctant to invest to this sector unless the person has a high degree of risk taking capacity. Moreover, like other sectors of the economy capital formation in agriculture also depends upon the profitability measured by the expected rate of return on investment. The condition is further accentuated where the traditional forms of capital predominate the sector because they are known to have low rate of return. Therefore, in a traditional agriculture the rate of investment and hence capital formation is very low. To transform traditional capital into modern capital, which generally yields higher rate of return, it is highly imperative to increase the rate of capital formation. The supply of capital is not the primary problem; it is rather a problem of determining the forms this investment must take, the forms that will make it profitable to invest in agriculture.[39] This would require to mobilise the resources and invest on those activities that will be profitable in the future. In traditional agriculture any strategy, which commands farmers to increase production, is doomed to failure even though they have access to knowledge. Instead an approach that provides higher incentives and rewards to farmers needs to be adopted.[40]

1.3 Source of Finance for Agricultural Capital Formation

The main sources of finance for agricultural capital formation[41] differ from country to country and from state to state. In a structurally backward economy like India agricultural capital formation is the sum total of capital formation accrued by public and private sector. By public sector we meant state government, non-departmental commercial undertakings, and local bodies. While private sector is consists of household and other private corporate sector. Those investments incurred by state government, non-departmental commercial undertakings, and local bodies for the formations of real asset are generally term as public capital formation. The public capital formation is mostly confined to the development and maintenance of infrastructures thereby creating a congenial environment for the farmers to invest on farm sector. While those investment made by the household and private corporate sector are termed as private capital formation. They are mostly confined to the creation of physical assets of the individual farms.

1.3.1. Public Capital Formation

In a structurally backward economy, public sector needs to play a crucial role in the development of agriculture sector. Public sector not only provide direct impetus to the growth and development of this sector, but also provides number of other support measure such as subsidization of key inputs, price support system of farm outputs, provision of institutional credits to the farmers etc. which would enlarge the productive base of the farm sector. The development of proper infrastructure coupled with facilities to access other support measures will induce the farmers to invest on farm sector.

Public capital formation in agriculture[42] is mainly confined to those areas where private individuals are reluctant to invest such as irrigation (major, medium and minor), crop husbandry, soil and water conservation, land reclamation, infrastructure, technology adoption and research.[43] These areas require a lumpy investment with long gestation period before the impulse of direct productive effects on investment is felt. At the same time return to investment in these areas are very low and takes long time in recovering the amount invested.

The public capital formation in agriculture is more of a rule rather than an exception the world over. The rationale of public investment is 'market failure' to provide public goods and is more common in the case of agricultural sector. The major attributes of such goods are indivisibility, externality, ignorance, jointness in supply and utilization.[44] As a result exclusive principle cannot be applied to benefits from such goods in the particular region where the goods are provided. Contrarily, exclusive principle can be applied in the case of private capital formation. In other words, public capital formation is subject to spill-over effect in the particular locality.

Public capital formation plays a very significant role in modernization of agricultural sector of an economy. An increase in public capital formation will enable to change the composition of capital suited to particular types of environment. It also enlarges the productive base of the agricultural sector. The success of green revolution in India was mainly due to massive public investment in basic infrastructure, electricity, irrigation, canals, roads and communications, organization of market structure, warehouses, research, extension services and number of other support measures.[45] Besides enhancing the production process directly public capital formation in agriculture is known to stimulate farmers' own investments in farm business and thereby resulting into further enlargement of the capital stock of this sector.[46] The extension of irrigation facilities and development of infrastructure leads to change in the cropping pattern and overall production technique as it creates a congenial condition for the application of new capital inputs such as chemical fertilizer, improved and high yielding varieties of seeds, pesticides etc.[47] In short, public capital formation provides impetus to the growth and development of agricultural sector by creating, improving and maintaining suitable infrastructure which will also creates conducive atmosphere for the farmers to invest on farm sector.

Agriculture being a state subject public capital formation has generally been treated as exogenous variable by micro-modellers. The overall public capital formation in agriculture is dependent upon the policy measures and availability of fund in the government exchequer. The implementation of suitable policy measures by the government will enable to acquire adequate public capital formation needed to remove those obstacles which impede the progress of the process of transformation of agriculture from a predominantly subsistence sector to a surplus generating business sector. It, however, requires a stable government with strong political will. The role of the government must evolve around those activities, which performed with the greatest effectiveness in terms of meeting the needs of the agricultural sector.[48] Instead of real investment if the government resort to the policy of appeasing the different interest groups and divert the available resources to current expenditures in the form of subsidies, exemptions, maintenance of existing projects and inordinate delays in completing the projects in hand, the growth and development of this sector will be delayed further. Over and above public investments should not meant to enrich the pockets of those concern politicians, bureaucrats and officials associated with.[49]

Amongst the various components of public capital formation in agriculture, investment in irrigation work accounted most. Therefore, any short run fluctuations in public capital formation were mainly due to the fluctuation in public investment in irrigation work and exert direct impact on the growth of output. However, in recent years lack of public investment in human capital formation seems to act as a limiting factor in the growth of agriculture. This is because the role of human capital formation in transforming the agricultural sector seems to be growing. In this modern era of complex pattern of production activities that characterize Japanese agriculture have been made possible by two types of public investment; investment in research and development and investment in general education.[50] Investment in these areas would lead to innovation and development of new capital and also increase the efficiency of agricultural labourers. It is with the help of these parts of capital formation that will enable to break those limiting factors acting upon the goal of modernising agriculture sector by altering the composition of material capital align to particular types of environment.

Public capital formation played a very crucial role in the development of agricultural sector. Any sort of fluctuation in public capital formation is bound to be a source of anxiety to the general public. Insufficient growth and receding total factor productivity would be due to inadequate public investment in agricultural research and extension, irrigation and other rural infrastructure, and diffused and inappropriate policy measures.[51] Adequate public capital formation along with suitable policy measures on the part of government sets the pace and pattern of agricultural growth in particular and economy in general. With the growing integration of world economy and declining state intervention market failure has come down considerably. Therefore, the rationale of provision of public capital formation in agricultural sector needs to be elucidated altogether.

1.3.2. Private Capital Formation

Private capital formation constitutes those reproducible assets and improvement in quality of existing assets, which are derived from investment made by household and corporate sector including both organized and unorganized segments. Private capital formation is mainly confined to the following areas; (i) Reclamation of land and land improvement, (ii) Purchase of new farm tools, and equipments including other means of transport, (iii) Construction and extension of irrigation means, (iv) New plantations and addition to existing orchards and plantations, (v) Construction of new farm structure such as farm houses, grain golas, cattle sheds, etc. (vi) Net increase in livestocks, (vii) Investment in farm inventories and other farm business.[52] The level of private capital formation would, however, be highly dependent upon the extent of infrastructural development which is the function of public capital formation in that region. Well-developed infrastructure will induce the farmer to invest on farm sector.

The main source of private capital formation in agriculture is household saving and borrowing. In the traditional agriculture the level of household saving and borrowing is very low due to the lack of adequate infrastructure to attract them. At the same time meagre saving available could not be utilized to an optimum level due to the lack of profitability, which hinges upon the expected rate of return. As a result the rate of private investment is very low and this sector continue to remain as subsistent sector. Profitability of investment is the key determinant of investment decision-making and is measured by the expected rate of return. It is generally perceived that if this return exceeds the rate of interest on funds borrowed for investment in question, the investment is a worthwhile proposition. However, unlike other sectors of the economy, due to the high degree of risk and uncertainty associated with this sector have rendered investment decision-making a truly complex process.[53] The problem is further accentuated in underdeveloped countries where traditional agriculture is prevalent. In these countries agricultural production is only at subsistence level. When farmers are assured of their expected return with certainty and investor has high degree of risk taking capacity private capital formation is bound to increase and vice versa.

Agriculture sector in every underdeveloped country is generally starved of investment resource because of low rate of household saving. It is the major source of finance in the case of private sector for the purpose of capital formation. In these countries private durable assets, which are produced in the industrial sector, using highly capital-intensive technique is very small. Usually in these countries private capital formation is mostly derived from unused family labour without much saving. In short, the incentive and ability to invest in underdeveloped countries are weak because the domestic market is small and domestic savings are meagre.[54] Nevertheless, private capital formation usually contributes a larger chunk of the total capital formation in agriculture sector of these countries. In India, it contributes nearly two-third of the total capital formation in agricultural sector. Private capital formation in agriculture has generally been treated as behavioral or endogenous variable as they are determined by a number of factors which differ over time and space. We shall discuss in detail this issue in Chapter V along with the issue on the relationship between public and private capital formation.

Given the importance of agriculture in Indian economy, the repercussion of a fall in agricultural growth will be felt in all sectors of the economy and more particularly the income and welfare of the poor who depend on agriculture.[55] However, the development of agriculture across the country is highly uneven. Baring few selected regions of the country green revolution seems to have bypassed most of the regions more particularly the states of the north eastern regions. Since India is a pluralistic country with varying landscape different region has got different cultural identity, mode of cultivation and institutional setup, therefore, a common policy programme may not be sufficient in achieving desire result across the region. In order to bring a balance regional growth, it is highly pertinent to study and analyse agriculture sector from different angle including capital formation for the country as whole and also for different regional settings. Such analysis would enable us to identify the major institutional and motivational obstacles, which stood on the way to modernization of this sector in different regional setting. It would also help the government to formulate appropriate policy measures instrumental for mobilizing the resources and canalizing them into more productive and socially desirable channels.

There are number of studies on agricultural capital

formation at the national level, as listed in bibliography, but hardly a few are available at the regional level. Though a number of studies and analysis on different aspects of agriculture sector of Manipur have been conducted so far; Islamuddin, Mohd. (1984); Singh, W. Kumar (1987); Singh, N. Ram (1987); Singh, N. Mohendro (1989); Singh, L. Bokul (1989); Singh, E. Bijoykumar (1992); Singh, Th. Biren (1993); the issue of capital formation in agriculture for the state of Manipur has not been analysed till today. Neither the Department of Economics and Statistics, Government of Manipur have initiated to quantify the capital formation in agriculture. In this era of modernization and declining trend of state intervention in the functioning of market, it has become all the more important to estimate and analyse various issues related with capital formation in agriculture sector both at the national level as well as for different regional setting. In the light of the above discussion on the relative importance of agricultural capital formation in achieving sustainable growth and development of this sector the present study endeavour to fill the void for the state of Manipur.

1.4 Objective of the Study

Estimates of capital formation, savings and their inter-sectoral variations go a long way towards providing for a better understanding of the economy and helping the government to formulate appropriate policy instruments for canalizing available savings into productive and socially desirable channels.[56] Dwelling upon this aspect present study is an attempt to quantify and analyse various issues related to capital formation in agricultural sector of Manipur. This study shall, however, be limited to tangible/material/ physical capital formation only. This is because while human capital formation is gaining wider acceptance, not much is known about the concept and their measurement except in education.[57] Further, we shall also skip the technological or engineering aspect of capital formation. The main objectives of present study are:

I. To construct a time series on capital formation in agriculture sector of Manipur.
II. To find out the determinants of capital formation in agriculture.
III. To find out the possible relationship between public and private capital formation in agriculture.
IV. To examine the effect of material capital formation in productivity and production of this sector.

1.5 Organisation of the Study

The present study has been organized into seven chapter including introduction. In chapter 2 a broad picture of agricultural sector in Manipur with special reference to agrarian structure and assets holding by the farmers has been discussed. Chapter 3 presents the sources of data, methodology followed and limitation in the construction of time series on capital formation in agriculture sector of Manipur. Chapter 4 presents the brief review on the various studies on the trend of public and private capital formation and causes and implications of these trends at the national level followed by the analysis on the same issue for the state of Manipur. A micro level analysis on the quantum of private capital formation has also been discussed in this chapter. The factors influencing the private capital formation both at the national and state level vis-à-vis complementarity hypothesis between public and private capital formation has been discussed in chapter 5. In chapter 6 an attempt has been made to examine the impact of capital formation in the production and growth of agriculture sector of

Manipur. It also reviews various studies on this issue at the nation level. Conclusion and policy implications of the study are presented in chapter 7.

NOTES

1. Bohm-Bawerk, E.V. (1948): *The Positive Theory of Capital*, William Smart (trans.), G. E. Stechert and Co., New York, p. 93.
2. *ibid.*, p. 96.
3. *ibid.*, p. 99.
4. Lal, R.N. (1977): *Capital Formation and its Financing in India*, Delhi School of Economics, Monograph in Economics, no. 3, Allied Publishers Pvt. Ltd., New Delhi, p. 1.
5. Tara Shukla (1965): *Capital Formation in Indian Agriculture*, 1st Edition, Vora & Co. Pvt. Ltd., Bombay, p. 2.
6. Lutz, F.A. (1961): "The Essentials of Capital Theory" in Lutz F.A. and D.C. Hague (eds.): *Theory of Capital*, New York, p. 16.
7. *ibid.*, p. 16.
8. Bohm-Bawerk, E.V. (1948): *op. cit.*, p. 38.
9. U.N.O. (1955): "Analysis and Projections of Economic Development", cited in M.L. Taneja and R. M. Myer: *Economics of Development and Planning (with special reference to India)*, Shoban Lal Nagin Chand & Co., Jalandhar, p. 325.
10. Clark, Colin (1957): *The Contentious of Economic Progress*, 3rd Edition, Macmillan & Co. Ltd., London, p. 566.
11. Tara Shukla (1965): *op. cit.*, p. 96.
12. *ibid.*, p. 6.
13. Nurkse, Ragnar (1980): *Problem of Capital Formation in Underdeveloped Countries*, 3rd impression, Oxford University Press, Delhi, p. 2.
14. *ibid.*, p. 2.
15. Alamjer, Mohiuddin (1976): "Rural Saving and Investment in Developing Countries: Some Conceptual and Empirical Issues", *Bangladesh Development Studies*, vol. 4, no. 1, p. 14.
16. Meier, G.M. (1970): *Leading Issues of Economic Development*, 2nd impression, Oxford University Press, New Delhi, p. 186.

17. Nurkse, Ragnar (1980): *op. cit.*, p. 31.
18. Meier, G.M. (1970): *op. cit.*, p. 600.
19. Lewis, W. Arthur (1971): "Economic Development with Unlimited Supplies of Labour" in A.N. Agarwala, and S.P. Singh (eds.): *The Economics of Underdevelopment*, Oxford University Press, Calcutta, p. 416.
20. Nurkse, Ragnar (1980): *op. cit.*, p. 8.
21. *ibid.*, p. 9.
22. *ibid.*, p. 9.
23. Government of India (1953): *First Five Year Plan*, Planning Commission, New Delhi, pp. 12-13.
24. The detailed analysis on each of these seven major contributions of agriculture to economic development is given in Gandhi, Vasant P. (1997): "The Role of Agriculture in India's Economic Development: A Survey", in B.M. Desai, (eds.): *Agricultural Development Paradigm for the Ninth Plan under New Economic Environment*, CMA Monograph No. 179, Oxford & IBH Publishing Company, New Delhi, pp. 107-121.
25. Todaro, Michael P. (1993): *Economic Development in the Third World*, First Indian Edition, Orient Longman Ltd., Hyderabad, p. 290.
26. Dandekar, V.M. (1986): "Agriculture, Employment and Poverty", Seminar paper presented at the conference organized by the Centre for Asian Development Studies, Boston University, October.
27. Todaro, Michael P. (1993): *op. cit.*, pp. 290-291.
28. The law of inheritance and succession in India permits the division and sub-division of one's property proportionately among one's successors. As the generation rolled on ancestral property also get divided and sub-divided thereby reducing the size of ownership holding. There is possibility of fragmentation of holding if the heirs demand a share in every plot of land. Hindu law of inheritance and succession amended in 1956 recognised daughter also as a full heir like son. Though the measure provides some sense of economic independence and equality of gender it is, however, economically not viable.

29. Sampath, R.K., et al. (1984): "Agricultural Production Functions, Technical Progress and Relative Land Utilisation of Large Farms Across Countries", *Indian Journal of Agricultural Economics*, vol. 39, no. 1, Jan-March, p. 53.
30. Schultz, T.W. (1964): *Transforming Traditional Agriculture*, Yale University Press, New Haven, USA, p. 107.
31. Panikar, P.G.K. (1969): "Capital Formation in Indian Agriculture", *Indian Journal of Agricultural Economics*, vol. 24, no. 4, Oct.-Dec., p. 146.
32. Ghosh, A. (2005): *Capital Formation and Entrepreneurship in Indian Agriculture*, Concepts Publishing Company, New Delhi, p. 1.
33. Schultz, T.W. (1964): *op. cit.*, p. viii.
34. *ibid.*, p. 146.
35. Todaro, Michael P. (1993): *op. cit.*, p. 296.
36. Schultz, T. W. (1964): *op. cit.*, p. 5.
37. Levi, John (1979): "Traditional Agricultural Capital Formation", *World Development*, vol. 7, nos. 11/12, Nov/Dec., p. 1054.
38. Schultz, T. W. (1964): *op. cit.*, p. 205.
39. *ibid.*, p. 4.
40. *ibid.*, p. 206.
41. The term 'capital formation' and 'investment' are interchangeably used here. The only difference between the two terms is that the former is the summation of fixed capital formation and change in stock while the latter term connotes only the fixed capital formation. It is mainly the fixed capital formation, which exerts influence on the production.
42. In recent years public capital formation has generated intense debate between the proponents of public capital formation in agriculture and public capital formation for agriculture. According to the later, CSO estimate is too narrow as it confined only to those activities evolved within the boundary of agricultural field. They viewed that public capital formation for agriculture which include all activities involve fully and partially to agriculture should be the appropriate measure

to be used while examining its impact on the growth and development of agricultural sector.

43. Banerjee, A. (1998): "Dynamic Capital Formation in Agriculture and Financial Reform", *Indian Journal of Agricultural Economics*, vol. 51, no. 4, Oct.-Dec., p. 567.
44. Desai, Bhupat M. and N.V. Namboodri (1997): "Government Expenditure on Agriculture under planning Era" in B.M. Desai (eds.): *Agricultural Development Paradigm for the Ninth Plan under New Economic Environment*, CMA Monograph No. 179, Oxford & IBH Publishing Company, New Delhi, p. 981.
45. Bhalla, G.S. (1977): *Changing Agrarian Structure in India: A Study of the Impact of Green Revolution in Haryana*, Meenakshi Prakashan, Delhi, p. 70.
46. Dhawan, B.D. (1998): *Studies in Agricultural Investments and Rural Savings*, Commonwealth Publishers, New Delhi, p. 114.
47. Ghosh, A. (2005): *op. cit.*, p. 12.
48. Nath, Vikas: "Trends in Public Investment in Agriculture", www.vikasnath.org, p. 3.
49. Dhawan, B.D. & S.S. Yadav (1997): "Public Investment in Indian Agriculture: Trends and Determinants", *Economic & Political Weekly*, vol. XXXII, no. 14, April 5-11, p. 710.
50. Schultz, T.W. (1964): *op. cit.*, p. 190.
51. Desai, Bhupat M. (1994): "Contribution of Institutional Credit, Self-finance and Technological Change in Agricultural Growth in India", *Indian Journal of Agricultural Economics*, vol. 49, no. 3, p. 465.
52. a. Reserve Bank of India (1978a): "Capital Formation of Rural Households During 1971-72", *All India Debt and Investment Survey*, p. 157.

    b. Alamgir, Mohiuddin (1976): *op. cit.*, p. 15.
53. Dhawan, B. D. (1996c): "Trends and Determinants of Capital Investments in Agriculture", *Indian Journal of Agricultural Economics*, vol. 51, no. 4, Oct.-Dec., p. 529.
54. Nurkse, Ragnar (1980): *op. cit.*, p. 5.
55. Nath, Vikas: *op. cit.*, p. 1.

56. National Council of Applied Economic Research (1979): *Capital Formation in Agriculture,* New Delhi, December, p. 13.
57. Schultz, T.W. (1961): "Investment in Human Capital", *The American Economic Review*, vol. 51, no. 1, March, pp. 1-17.

2

Agrarian Structure of Manipur

2.1 Introduction

Manipur, "a jewel of India" as aptly remarked by Pandit Jawaharlal Nehru, is situated on the eastern frontier of India. It is bounded on the north by Nagaland, on the south partly by Myanmar and partly by Mizoram, on the east by Myanmar and on the west by Assam. Manipur lies between 93.03° E and 94.78° E longitude and between 23.83° N and 25.68° N latitude.[1] The total geographical area of the state is 22,327 sq. km. which constitute about 0.7 per cent of area of the Indian Union. Physically Manipur may be divided into two parts - the hills and the valley. The valley area is at the center surrounded by hill ranges which are considered to be the off-shoots of the Himalayan ranges. The valley area, which consists of four districts viz. Imphal West, Imphal East, Thoubal and Bishnupur, constitutes only 2,238 sq. km. and is just over 10 percent of the total area of the state. The rest nine-tenth of the state is hills. However, the four valley districts support 61.54 percent of total population of the state. It shows that valley areas are densely populated as compared to hill districts. The density of population in the valley is 631 persons per sq. km. as against 44 persons per sq. km. in the hills as per 2001 census. The overall density of population of the state is 103 persons per sq. km.

TABLE 2.1

District-wise Distribution of Area, Population and Density of Population of Manipur (2001 Census)

District	Area in sq. km.	Population	Density of Population (Persons per sq. km.)
A) Valley	2,238	14,11,766	631
Imphal East	709	3,94,876	557
Imphal West	519	4,44,382	856
Bishnupur	496	2,08,368	420
Thoubal	514	3,64,140	708
B) Hills	20,089	8,82,130	44
Senapati	3,271	2,83,621	87
Churachandpur	4,570	2,27,905	50
Chandel	3,313	1,18,327	36
Ukhrul	4,544	1,40,778	31
Tamenglong	4,391	1,11,499	25
Manipur	22,327	22,93,896	103

Source: 1. Population: Census of India 2001, Series 14, Manipur, Paper 1 of 2001.

Note: 1. Excludes Moa-Maram, Paomata and Purul sub-division of Senapati district of Manipur.

2. Imphal district has been bifurcated into Imphal West and Imphal East in 1997.

3. In most of the recent publications of the state government the total area of the state is 22,327 sq. km. The Report of the working group on Development of North-Eastern Region during the Seventh Five Year Plan, NEC (1985) placed it at 22,360 sq. km. which is closer to the figure of 22,356 sq. km. given in the statistical Handbook of Manipur, 1978.

The altitude in the valley area of the state is 790 metres above mean sea level. The hill ranges are higher on the north and gradually diminish in height as they reach the southern part of the state. Similarly, valley areas also slope towards the south. Soils in the hills are derived mostly from shale and sandstone and are mostly lateritic in nature. The soils are invariably acidic in reaction with high organic matter and low phosphate content. The texture of the soil varies from clayey to fine. The soils of the valley areas are derived from shale and siltstones. These soils are acidic in reaction with humus content. Phosphate content in the valley is slightly higher than that of the hill areas. The texture of the soil varies from fine to very fine.[2]

In an analysis of soil samples for PH organic carbon and exchangeable potassium, calcium and magnesium in 8 blocks from valley region, it was found that the soil was acidic in nature with a PH range of 4.8 to 6.65. Organic carbon content was high in most of the samples ranging from 0.82 to 1.94%.[3]

According to the Techno-Economic Survey of Manipur the soil in the valley is very fertile, enriched, as it is, by the silt deposit brought and spread by the rivers, which overflow their banks regularly. While potash and lime ingredients are adequate, it is deficient in nitrogenous and phosphoric contents. The hill soils ranges from laterite to sandy loam and are deficient in nitrogenous and phosphoric contents.[4]

Manipur has a salubrious climatic condition, which varies with location and elevation. The central valley area of the state has sub-tropical climate whereas in the surrounding hills, it varies from semi-temperate to temperate depending on the altitudes of the hill ranges. The hill areas are cooler and wetter than the valley areas.

The average annual rainfall in the Imphal valley areas is about 1030 mm. The distribution of rainfall in the hill ranges also varies with location and elevation. The rainy season in Manipur starts from May and continues till October when the state is under the spell of the southwest monsoon.

2.2 Agriculture in Manipur

Agriculture occupies the pride of place in the economy of Manipur. It is the backbone of the state economy and mainstay of the people of Manipur. Though the share of agriculture in state NSDP has declined over the years the proportion of population employed continued to remain high. It contributes over 33% of the Net State Domestic Product and provides employment to about 52 percent of the total working population of the state according to 2001 census. Apart from direct contribution, a high correlation coefficient of 0.8015 during the period (1972 to 1997) studied between real NSDP originating from agriculture and from non-agriculture indicates that there is a significant association with other sectors of the economy. Agriculture being the main contributing sector to state income (NSDP), any variation in the production and productivity of this sector would eventually lead to fluctuation in the state's income and more particularly the income and welfare of the people who depend on it. Despite being the leading contributing sector in the state economy, some of signs of the traditional agriculture, which we have inherited from our ancestors, are still woven around this sector in spite of 50 years of economic planning. Therefore, this sector needs to be studied in depth so as to identify the institutional and motivational obstacles which stand on the way to modernization of this sector.

Since there has been a clear distinction of topographical

landscape of Manipur into valley and hills, the development of agriculture across the state is also highly uneven. This can be seen when we look on the cultivation of cereal crops in the state given in Table 2.2.

TABLE 2.2

District-wise Area under Cereal Crops for the Agricultural Year 1999-2000 in Manipur

(Area in '000 hect.)

District	*Total area*	*Area under cereal crops*	*Percentage of area under cereal crops*
A) Valley	223.8	89.24	39.87
Imphal East	70.9	33.65	47.46
Imphal West	51.9	15.46	29.79
Bishnupur	49.6	18.36	37.02
Thoubal	51.4	21.77	42.35
B) Hills	2008.9	72.15	3.59
Senapati	327.1	26.77	8.18
Tamenglong	439.1	9.55	2.17
Churachandpur	457.0	14.57	3.19
Chandel	331.3	6.35	1.92
Ukhrul	454.4	14.91	3.28
Manipur	2232.7	161.39	7.23

Source: Government of Manipur, "Crop Estimation Survey in Manipur, 1999-2000", Directorate of Economics and Statistics.

The area under cereal crops in the valley area constitutes 39.87 percent of the total valley area. Amongst the valley districts, Imphal East has the highest percentage of area under cereal crops, which accounts 47.46 percent of the area under this district. In the hills, however, only 3.59 percent of the total area has been utilized for cultivation of cereal crops. Amongst the hill districts, Senapati has the highest percentage with 8.18 percent and the lowest is

Chandel with 1.92 percent. The low percentage of cereal cultivation in the hill districts is mainly due to the undulating hill ranges and lack of proper infrastructures.

Rice is the staple food crop of Manipur. Out of the total area under cereal cultivation over 91 percent in 1972-73 and 97 percent in 1999-2000 were under rice cultivation. Similarly, over 87 percent in 1972-73 and 96 percent of the total cereal production in the state in 1999-2000 were rice. An examination on the performance of rice cultivation in the state can, therefore, be summarized as the extent of agricultural development in the state. .

Due to the varied topography, the development of rice cultivation in the state also varies accordingly. The valley area is aptly known as the 'rice bowl' of the state. In 1999-2000, the valley area which constitute 10 percent of the total geographical area of the state, accounts 56.82 percent of the total area under rice cultivation and 64.98 percent of the rice production in the state. The rest is accounted from the surrounding hill regions which constitute 90 percent of the total geographical area of the state.

The type of rice cultivation in the state also varies according to the location and elevation of the region. In the valley areas, cultivation is of permanent type whereas in the hills shifting/jhum cultivation is the predominant mode of cultivation. Till 1984-85, due to the lack of cadastral survey no precise area of land under shifting/jhum cultivation in the state has been identified. However, for the first time in 1985-86, Directorate of Economics and Statistics, Government of Manipur has taken up the responsibility of measuring the area under jhum cultivation. Since then it has been conducting the survey on area under jhum cultivation annually and published in annual Crop Estimation Survey in Manipur.

The area reported under jhum cultivation in 1985-86 was 25.23 thousand hectares and remained relatively stable till the end of 80s. It then declined to 23.81 thousand hectares in 1990-91. Thereafter it has been increasing continuously which reached to 40.54 thousand hectares in 1999-2000. In terms of proportion, with respect to the total area under rice cultivation in the hill districts, it has been increasing continuously. The percentage of area under jhum cultivation in the hills has increased from 37.97 percent in 1985-86 to 59.73 percent in 1999-2000. The trend on the area under permanent and jhum cultivation, since 1985-86, is given in Appendix 2.1. The whole area under jhum cultivation is reported only in the hill districts of Manipur.

TABLE 2.3

District-wise Area under Rice Crop and by Types of Cultivation in Manipur in 1999-2000

(Area in '000 hect.)

District	*Area under PMT*	*Area under Jhum*	*Total area*	*Percentage area under Jhum to total area under rice cultivation*
A) Valley	89.24	--	--	--
Imphal East	33.65	--	--	--
Imphal West	15.46	--	--	--
Bishnupur	18.36	--	--	--
Thoubal	21.77	--	--	--
B) Hills	27.27	40.54	67.81	59.78
Senapati	22.66	1.87	24.53	7.62
Tamenglong	1.07	8.47	9.54	88.78
Churachandpur	1.72	12.01	13.08	91.82
Chandel	1.10	5.07	6.17	82.17
Ukhrul	0.72	13.12	13.84	94.80
Manipur	116.51	40.54	157.05	25.81

Source: Government of Manipur, "Crop Estimation Survey in Manipur, 1999-2000", Directorate of Economics and Statistics.

Table 2.3 shows the district-wise distribution of area under rice crop by different types of cultivation in 1999-2000. In the hills, out of 67.81 thousand hectares under rice cultivation, 40.54 thousand hectares have been put under jhum cultivation, which constitute about 60 percent of the area under rice cultivation. Amongst the hill districts, most extensive shifting cultivation is practiced in Ukhrul district where 94.8 percent of the total area under rice cultivation in the region was put to this mode of cultivation, followed by Churachandpur district with 91.82 percent. It is least practiced in Senapati district with only 11.39 percent. In the valley area this mode of cultivation is not at all reported. While the area under permanent (including terrace) types of cultivation in the state declined over the year, area under jhum cultivation in the hills has been increasing rapidly particularly in the 90's. The declining trend of area under permanent type of cultivation in the state may be the result of increased in population and subsequent urbanization. This implies that the arable land in the state is under tremendous pressure in accommodating and filling the mouth of ever burgeoning population in the state.

On the other hand, rise in area under jhum cultivation in the state is a major source of concern, as it would lead to severe environmental degradation and ecological imbalance in the state. However, jhum cultivation in the state is deeply rooted in the culture and tradition of the tribal community. It has a special significance in the ethos of the tribal society and social relationships, cultural values and mythical belief.[5] It, therefore, cannot be restricted and abandoned immediately. Instead, an attempt should be made to modify the present mode of jhum cultivation, similar to that of NEPED[6] in Nagaland, so that it is economically viable, socially acceptable and ecologically sustainable. The combination of eco-friendly jhum cultivation along with the implementation of bio-village approach[7] can achieve an appropriate balance between conservation and commercialization, traditional wisdom and modern technologies.[8]

2.2.1 Cost of Cultivation of Rice Crop in Manipur

In the following section we shall examine the change in the cost structure of rice cultivation in the valley areas of the state, based on the information provided by the Department of Agriculture, Government of Manipur which is given in Appendix 2.2 (A & B), between the 80's and the 90's. The summary of cost of cultivation of one hectare of HYV and local variety of paddy during the 80's and the 90's are shown in Table 2.4.

TABLE 2.4

Share in Cost of Cultivation of Rice (HYV and Local Paddy per hectare) in the Valley Areas of the State during 80's and 90's

(in percent)

	HYV		*Local*	
	80's	*90's*	*80's*	*90's*
Mandays required	77	83.3	83	88
Fertilisers	13.7	11.3	7.8	6.5
Seeds	4.2	2.7	4.8	3.1
P.P. Chemicals	5.1	2.7	4.4	2.4

Source: 1. Cost of Cultivation data provided by the Directorate of Agriculture, Government of Manipur.

2. Singh, E.B.K. (1992): *op. cit.*, pp. 407-411.

The above table shows the following features:

1. Even though labour's share in cost of cultivation of HYV paddy has increased between the 80's and

the 90's it is still lower than that of local paddy. The gap, however, has narrowed down from 6 percent to 4.7 percent. This is mainly due to the increase in the labour man-days required in the cultivation of HYV paddy from 244 in the 80's to 271 in the 90's, which is about 11.06 percent. As against this for local paddy, it increased from 230 to 247, which is about 7.4 percent during the corresponding period.

2. The share of fertiliser and plant protection chemical in the cost of cultivation of HYV paddy continued to be higher than that of local variety even though their shares have declined in the 90's.
3. Though the share of seed in the cultivation of HYV paddy has declined in the 90's it is still lower than that of local variety.

Further, the total value added in the cultivation of HYV and local paddy during the 80's and 90's is decomposed to examine the shares of land, labour, and capital/profit are given in Table 2.5.

The table reveals the following characteristics:

1. The labour's share in HYV paddy cultivation has declined marginally while that of local paddy has increased by 7.4 percent thereby widening the gap between the two.
2. The share of land has declined by 3.1 percent for HYV paddy while that of local paddy, it increased by 9.5 percent thereby leading to widening of gap between the two varieties.
3. While the capital/profit's share in the total value added of HYV cultivation has increased by 3.4 percent that of local paddy is declined by 17.9 percent.

TABLE 2.5

Cost Structure for HYV and Local Paddy Per hectare in the Valley Areas of Manipur in the 80's and the 90's

	HYV		*Local*	
	80's	*90's*	*80's*	*90's*
Total wage cost	Rs 3025.60	Rs. 12,100.15	Rs. 2852.00	Rs 11,028.55
Intermediate input cost	Rs. 905.00	Rs. 2,428.00	Rs. 585.00	Rs. 1,510.00
Gross value of output	Rs. 16,000.00	Rs. 65,000.00	Rs. 12,000.00	Rs. 39,000.00
Value added	Rs. 12,069.40	Rs. 50,471.85	Rs. 8,563.00	Rs. 26,461.45
Rent	Rs. 4,000.00	Rs. 15,522.00	Rs. 3,200.00	Rs. 12,415.00
Labour's share in value added	25.1 percent	24.0 percent	33.3percent	41.7 percent
Land's share in value added	33.1 percent	30.8 percent	37.4 percent	46.9 percent
Capital/profit share	41.8 percent	45.2 percent	29.3 percent	11.4 percent

Sources and Assumption: For 90's

1. Wage cost and intermediate input cost have been derived from the cost of cultivation data provided by the Directorate of Agriculture, Government of Manipur.
2. Yield per hectare has also been given as 5000 kgs for HYV paddy and 3000 kgs for local variety.
3. Price of rice has been considered as Rs. 13/kg for both HYV and local variety.
4. Rent per hectare is 1194 kgs which corresponds to 20 pots per hectare for HYV paddy and 955kgs or 16 pots per hectare for local paddy.
5. For 80's see Singh, E. B. K. (1992): *op. cit.*, pp. 407-414.

The following features revealed that in the 90's the labour used in the cultivation of HYV and local paddy in the state has increased. On the other hand, the intermediate input used has declined. Thus, we can conclude that the cultivation of rice crop in the state is more labour intensive in the 90's than in the 80's. The cultivation of HYV paddy is more profitable than that of local paddy. This is mainly due to the increase in the yield of HYV paddy and decline in the yield of local paddy while their prices remained same at Rs.13 per kg.

2.2.2 Growth of Rice Production in Manipur

A sustainable growth in agricultural production and productivity is very much essential for overall stability of the state economy. It, therefore, is highly imperative to examine the performance of rice cultivation in the state. One way of measuring the performance of rice cultivation in the state during 1972-99 is to calculate the annual compound growth rate of area, production and yield of rice whose raw data are given in Appendix 2.3. There are various models by which the annual compound growth rate can be calculated for a specified data series.[9] However in our present study we shall estimate the annual compound growth rate of area, production and yield by using log-linear model;

$$\log Y = a + bt$$

Where Y is area/production/yield of rice and t represents time variable. The growth rate is estimated by $(e^b - 1) \times 100$. This model is fitted to original time series data, three and five yearly moving average. The estimated annual compound growth rate of area, production and yield of rice crop in the state during the period 1972-73 to 1999-2000 is given in Table 2.6.

In all three data series viz. original, three and five yearly moving average for area, production and yield of rice fitted into the model, the coefficients of t are found to be statistically significant at 1 & 5 percent level of significance. If the criteria for selecting the model having highest $\bar{R}^2$ has been chosen, it has been found that in all three - area, production and yield of rice in the state, the compound growth rate calculated from five yearly moving average is the most appropriate one.

TABLE 2.6

Compound Growth Rates of Area, Production and Yield of Rice in the State during 1972-99

	Types of series	*Estimates of b*	$\bar{R}^2$	*r*
Area				
	Original Time Series	- 0.0031108 (- 2.163)**	0.1199	-0.3106
	Three Yearly Moving Average	- 0.0042652 (-6.555)*	0.6267	-0.4256
	Five Yearly Moving Average	- 0.0048312 (-12.957)*	0.8789	-0.482
Production				
	Original Time Series	0.018109 (5.442)*	0.5145	1.8274
	Three Yearly Moving Average	0.01535 (8.625)*	0.7459	1.5468
	Five Yearly Moving Average	0.014992 (9.429)*	0.7926	1.5105
Yield				
	Original Time Series	0.021219 (7.905)*	0.6949	2.1446
	Three Yearly Moving Average	0.01982 (11.864)*	0.8483	2.0018
	Five Yearly Moving Average	0.020054 (13.363)*	0.8853	2.0256

Note: 1. * & ** Significant at 1% & 5% level of significant

2. Figures in the parenthesis indicate t statistics.

Further, the test of acceleration or deceleration in the estimated trend growth rates has also been conducted. The hypothesis of acceleration or deceleration in the estimated trend growth rates is examined by using the modified exponential;

$$\text{Log } Y = a + bt + ct^2$$

The summary results of the test are given in Table 2.7.

TABLE 2.7

Rate of Change of Growth Rates of Area, Production and Yield of Rice in the State

	Types of series	*Estimates of b*	*Estimates of c*	*Remark*
Area				
	Original Time Series	-0.00017033 (-0.028)	-0.00010139 (-0.501)	
	Three Yearly Moving Average	-0.005258 (1.909)	0.000036773 (0.371)	
	Five Yearly Moving Average	-0.0049065 (-3.083)	0.000003012 (0.049)	
Production				
	Original Time Series	0.016155 (1.149)	0.000067369 (0.143)	
	Three Yearly Moving Average	-0.00018505 (-0.027)	0.00057538 (2.361)**	Accelerating
	Five Yearly Moving Average	-0.0026475 (-0.0481)	0.00070557 (3.299)*	Accelerating
Yield				
	Original Time Series	0.016312 (1.444)	0.00016923 (0.448)	
	Three Yearly Moving Average	0.0054076 (0.847)	0.00053379 (2.327)**	Accelerating
	Five Yearly Moving Average	0.0024111 (0.479)	0.00070571 (3.613)*	Accelerating

Note: 1. * & ** Significant at 1% & 5% level of Significant respectively
2. Figures in the parenthesis indicate t statistics.

The compound growth rate of area under rice cultivation is -0.48 percent per annum and it neither accelerates nor decelerates. On the other hand, production and yield of rice grows at 1.51 percent and 2.03 percent per annum respectively and are both accelerating. The positive with accelerated growth of rice output in the state during 1972-99 is the result of highly positive accelerated growth rate of yield, which offset neither accelerating nor decelerating negative growth rate of area under rice. It indicates that the present technique of rice production in the state is land saving. This can be further endorsed by the decomposition result of growth of rice output in the state during the corresponding period.

The decomposition of growth of output into different component parts so as to identify the major contributing component of the growth of output is another way of looking at the growth of output. Different models are in use in this regard.

Minhas and Vaidyanathan (1965) decomposed the growth of crop output in India for the period 1951-61 into four components viz. area effect, yield effect, cropping pattern effect and interaction between yield and cropping pattern effect.[10] Dharam Narain (1976) decomposed the growth of productivity into cropping pattern effect, locational shift effect and yield effect.[11] The interaction of cropping pattern effect and locational effect are also considered. Desai and Nambooderi (1997) decomposed the growth of land productivity into pure yield effect, cropping pattern effect, location effect and interaction effects between cropping pattern and yield and location and yield effect.[12] Singh (1992) decomposed growth of rice output in Manipur for the period 1965-87 into area effect and yield effect.[13] But the cropping pattern effect and interactions effects were not considered.

For decomposition of rice output in Manipur, we start from the following identity $Q_t = A_t \times Y_t$, where Q_t, A_t, and Y_t indicates output, area, and yield in period t.

The additive scheme of decomposition is as follows:

$$Q_t - Q_0 = (A_t - A_0)\, Y_0 + [(Q_t - Q_0) - (A_t - A_0)Y_0]$$

Where, $(A_t - A_0)\, Y_0$ = Area effect

$[(Q_t - Q_0) - (A_t - A_0)\, Y_0]$ = Yield effect

The shares of area expansion and yield improvement in the total increase in output when expressed as percentages give the area effect and yield effect respectively.

In the multiplicative scheme of decomposition it is assumed that Q, A, and Y are growing at the rate of r, a, and y respectively.

We than write[14]

$$Q_t = Q_0\,(1 + r)^t = A_0\,(1+a)^t\, Y_0\,(1 + y)^t$$

Where $r = \left(\frac{Q_t}{Q_0}\right)^{\frac{1}{t}} - 1$ = Rate of growth of rice output.

$a = \left(\frac{A_t}{A_0}\right)^{\frac{1}{t}} - 1$ = Rate of growth of area under rice.

$y = \left(\frac{Y_t}{Y_0}\right)^{\frac{1}{t}} - 1$ = Rate of growth of rice yield.

Taking logarithm to above equation and simplifying, then taking antilog, one can write;

$$r = a + y + ay$$

$$r \cong a + y$$

In both the schemes we exclude the effect of cropping pattern and effect of interaction.

TABLE 2.8

Result of Decomposition Exercise for Rice Output in Manipur (1972-1973 to 1999-2000)

Model	*Additive Model*		*Multiplicative Model*	
	Area Effect	*Yield Effect*	*Area Effect*	*Yield Effect*
Original	5.10	94.90	7.89	92.11
Three Yearly Moving Average	-3.44	103.44	- 4.5	104.5
Five Yearly Moving Average	-14.88	114.88	-18.97	118.97

Both the additive and multiplicative models confirmed the overwhelming importance of the yield effect in the growth of rice output. In fact the area effect is found to have negative impact on the growth of rice output for the period of the moving averages. The result of decomposition exercise of growth of rice output for the period 1965-87 shows that 42 percent of the growth of rice output in the original data, 18 percent in the three yearly moving average and 28 percent in the five yearly moving average time series can be attributed to area effect in multiplicative model. In the case of additive model the corresponding figures are 37.6 percent, 17.06 percent and 26.24 percent respectively.[15] However, in the present study covering 1972-99 the area effect in both models for three yearly and five yearly moving averages is found to have negative impact on growth of rice output. The shares of yield effect and area effect, however, are found to be differ with the period of moving averages. The fluctuation in the share of effects may be partly attributed to the presence of cyclical component in the time series.

Thus, we can conclude that area under rice cultivation in the state have declined while the per hectare yield of rice output is increasing. In other words, while the area growth rate neither accelerates nor decelerates, yield growth rate is accelerating thereby resulting into accelerated growth rate of rice production. Not only is the growth of area under rice cultivation declining, the share of area in the growth rate of rice output in the state is negative. Hence the scenario of accelerating yield rate together with a dominant share of rate of growth of yield in growth rate of rice output confirms the importance of capital formation via HYV based technological change in the agricultural sector of Manipur.

2.3 Land Use Pattern

Land is the most precious gift of nature. With rising pressure of population on land, its supply remaining the same, proper utilization of this precious gift of nature should always be given due importance. It is, therefore, highly imperative to study the existing land use pattern in the state.

Land utilization statistics provide the area figure showing distribution of the total geographical area of the country/state into its various uses.[16] In other words, it shows in detail the distribution of existing land according to its actual use and not how a particular piece of land may be utilized. The existing land use pattern in different regions in India has been evolved as the result of interaction of various factors, such as the physical characteristics of land, the institutional framework, the structure of other resources available like capital, labour, etc. and the location of the region in relation to other aspects of economic development such as those relating to transport, as well as to industry and trade. The present pattern can therefore be viewed as a static harmony and adjustment with the main characteristics of the economy of the region.[17]

Till 1949-50, the land utilization statistics were available under five categories, namely;

1. Forest,
2. Area not available for cultivation,
3. Other uncultivated land excluding current fallows,
4. Fallow land, and
5. Net area sown.

However, this classification was found to be quite inadequate to meet the needs of agricultural planning. Therefore, from 1950-51 onwards the present nine-fold classification was introduced. They were;

1. Forest,
2. Land put to non-agricultural uses,
3. Barren and uncultivable land,
4. Permanent pastures and other grazing lands,
5. Miscellaneous tree crops and groves not included in the net area sown,
6. Cultivable waste,
7. Fallow land other than current fallows,
8. Current fallows,
9. Net area sown.

Besides these nine categories two more categories viz. (i) area sown more than once, and (ii) total crop area, were also usually worked out for additional information.

The total of these nine categories add up to the reporting area. Categories 7, 8, and 9 together gives total arable land. When categories 4, 5, and 6 are added to total arable, we have potential land available for crop husbandry.

TABLE 2.9

Land Use Pattern in Manipur 1992-93 (Provisional)

Sl. No.	*Class*	*Area in '000 hect*	*Percentage of reporting area*
1.	Reporting area	2,211	100
2.	Forest	602	27.23
3.	Land put to non-agricultural uses	26	1.18
4.	Barren and uncultivable land	1,419	64.18
5.	Permanent Pastures land other grazing land	n	--
6.	Land under misc. tree crops and groves not included in net area sown	24	1.08
7.	Cultivable wasteland	n	--
8.	Fallow land other than current fallows	--	--
9.	Current fallows	a	--
10.	Net area sown	140	6.33
11.	Area sown more than once	47	2.13
12.	Total cropped area	187*	8.46

Source: "Agriculture in Brief", 1998 issue, 26th Edition, Directorate of Economics and Statistics, Department of Agriculture and Cooperation, Ministry of Agriculture, Government of India, Pp. 226-229.

Note:-n = included under the head "land under misc. tree crops and groves not included in net area sown.

a = Below 500 hectares.

* = In the absence of data for the year 1992-93, latest available information has been utilized.

Table 2.9 shows that the reporting area in Manipur is 2211 thousand hectares. However, out of this total reporting area only 4.54 percent is based on complete enumeration and rest is based on ad-hoc estimation. The corresponding figures for agriculturally advanced states liked Punjab and Haryana are 100 percent based on complete enumeration.

For Manipur, area of arable land for crop husbandry is 140 thousand hectares, which gives a 100 percent exploitation of arable land. Potential land available for crop husbandry is 164 thousand hectares, which gives 85.37 percent exploitation of potential land.[18] It also shows that, as much as 64 percent of total reporting area of Manipur are classified under 'Barren and unculturable land'. This category includes hilly areas, which cannot be brought under cultivation except at a high cost whether such land is in isolated blocks or within cultivated holdings. Land under forests accounted for 27.23 percent of the total reporting area. However, the proposed percentage that has to be maintained according to National Forest Policy Resolution, 1952 was 33.3 percent, which is to be revised to 60 percent in the hill regions and 20 percent in the valley. Since the proportion of area under forests is already lower than the proposed level, the scope of expanding net area sown by claiming area under forests for crop husbandry is very much limited.

2.4 Agrarian Structure

The agrarian structure of a country is a historically evolved complex of customs, institutions, rural power configurations and a number of factors that shape the relations of productions and govern the distribution of agricultural output among various groups and classes involved in agricultural production. In short, it refers to the manner in which man-land relationships are governed. It covers the pattern of composition of different groups of individuals in the agricultural community in terms of land ownership, land tenurship and land holdings. In other words, the agrarian structure of a country illustrates the way in which land is held and cultivated and rights and privileges enjoyed by the different categories of people.

However, there is no suitable single empirical variable to represent the various facets of agrarian structure. Any study on capital formation in agricultural sector of an economy without a proper analysis of the prevailing agrarian structure would be incomplete.

In the following sections we shall examine the following aspects of the agrarian structure of Manipur.

1. Pattern of distribution of land
2. Distribution of assets

In the absence of sufficiently long time series data the cross section perspective of the state will be relied upon.

2.4.1 Distribution of Operational Holding

Data in this connection have been collected through the various Agricultural Census and sample surveys conducted by the National Sample Survey Organization.

2.4.1.1 Agricultural Census Data

The first ever world Agricultural Census was conducted in 1930 under the guidance of the International Institute of Agriculture. From the second round onwards world Census of Agriculture was conducted under the guidance of the Food and Agricultural Organization (FAO) of the United Nation. India could not participate fully up to the 3rd round due to lack of adequate resources and various other reasons. However data required by Food and Agricultural Organization (FAO) were collected from second round onwards through the 8th Round of NSS (July 1954-April 1955). In 1960, the NSS collected data on land holdings in its 16th (July 1960-August 1961) and 17th (September 1961-July 1962) rounds with agricultural year 1959-60 as the reference period.[19]

The first comprehensive agricultural census of India

was carried out on complete enumeration basis in 1970-71 coinciding with the fourth round of world Agricultural Census. Since then the subsequent censuses were conducted for every five years with greater scope and coverage. In major parts of India comprehensive land records exist containing plot-to-plot details under well-established revenue administration system. In these areas census data were collected on complete enumeration basis by re-tabulating the information available in the land records. The basic unit of enumeration is the operational holding.

An operational holding is defined as all land which is used wholly or partly for agricultural production and is operated as one technical unit by one person alone or with others or without record to the title, legal form, size or location. The term 'technical unit' used in the definition is defined as that unit which is under the same management and has the same means of production such as labour force, machinery and animals. Thus it is the actual cultivator and not the owner of the land that has been used as basic unit of data collected in the agricultural census. Data on the following items were collected with operational holding as the basic unit;

1. Number of operational holding by size class,
2. Area under crops,
3. Land utilization,
4. Irrigation - source-wise and crop-wise,
5. Tenure and Tenancy.

In the remaining states and union territories including Manipur where comprehensive land records do not exist the required data were collected through sample survey. In these states/UTs data on agricultural inputs like use of fertilisers and pesticides, machinery and implements, livestock owned etc. were also collected by sample survey.

The land record system that has been prevailing in Manipur is highly inadequate mainly because of insufficient fund available to the Revenue Department for maintenance and updating of land records and for improvement and modernization. In the absence of required revenue agency for collection of data and cadastral survey on all cultivated land, all the censuses held so far in Manipur were conducted through sample survey. A comparative picture of land ceiling in Manipur is given below.

TABLE 2.10

Land Ceiling in Manipur

(in hectare)

Land ceiling	*Irrigated with 2-crops*	*Irrigated with 1-crop*	*Dry land*
Suggested in National Guidelines of 1972	4.05 to 7.28	10.93	21.85
Suggested lower ceiling	3.0	7.5	12
Actual ceiling	5.0	5.0	6.0

Source: 1. Government of India: Annual Report 1986-87, Ministry of Agriculture, Department of Rural Development.

2. Government of India: Agricultural Statistics At-a-Glance, April 2000, Directorate of Economics and Statistics, Department of Agriculture and Cooperation, Ministry of Agriculture.

In the Agriculture Census 1970-71, all subdivision of the central district viz. Imphal East, Imphal West, Thoubal, Bishnupur, Jiribam, Tengnoupal, Chandel, Chakpikarong and 13 selected villages of Churachandpur subdivision, south district were covered. Enumerators went to each village and enlisted the entire operational holdings of the village following as far as possible in the 1971 census serials. For detailed enquiry 5 percent sample was taken from each village. Detailed information collection was completed by mid 1973.

The items covered in 1970-71 census are as follows:

1. Number of operational holdings and area operated,
2. Number and area of holdings by tenure,
3. Area under different land uses
4. Number of holdings reporting irrigation and area irrigated,
5. Source-wise area irrigated,
6. Area under various crops,
7. Information on fertilisers,
8. Inventory of livestocks.[20]

The agricultural census 1970-71 and other later censuses are not comparable because in 1970-71 census only the subdivisions of central district and 13 selected villages of south district were covered. Whereas the later series of censuses conducted so far covered all the districts of Manipur.

From second census onwards agricultural censuses were conducted in two phases viz. Main Census and Input Survey. In the main census, a stratified two-stage sampling design with the revenue village as the first stage sampling unit and operational holdings as the second stage sampling unit were adopted. The whole state was divided into 25 sub-division/tehsil and each one of them is constituted a stratum. From each stratum, at least 20 percent of the villages were selected by simple random sampling method. In each selected village, all the operational holdings were enumerated by enquiry. Information on the following items were collected;

1. Number and area of operational holdings in size class-wise,
2. Tenancy particulars,
3. Land utilisation,
4. Area under crops, and
5. Crop-wise and source-wise area irrigated.

In the input survey, a sample of 125 villages was selected from central district through the circular systematic random selection and was listed holding-wise based on the residential cultivators. 20 holdings were selected from each village by taking 4 holdings from each size class of 5 group of the input data viz., (1) Number of parcels, (2) Multiple cropping, (3) Application of chemical fertilisers, manures and pesticides, (4) Livestocks, (5) Agricultural machinery and inputs.[21]

2.4.1.2 National Sample Survey Data (NSS)

National Sample Survey Organization (NSSO) has conducted 5 sample surveys so far with a view to obtain quantitative information on the agrarian structure of the country, which is relevant to land policy.[22] The first survey on land holdings was conducted in the eighth round (July '54 - April '55) as part of the 1950 programme of the world Agricultural Census initiated by the Food and Agricultural Organization (FAO) of the United Nations. In the survey, information on operational and ownership holdings were collected. The second survey was conducted in the 16th (July '60-August '61) and 17th (September '61-July '62) rounds of the NSS coinciding with the world Agricultural Census Programme of 1960.

Since then, NSS has been regularly conducting survey on land holdings for every ten years with some modifications in item coverage and the method of data collected. The third Land Holdings Survey was conducted in its 26th round (July '71-June '72) and fourth in 37th round (January-December '82). The fifth survey in the series was conducted in the 48th round (January-December '92).

Information on various aspects of ownership and operational holdings were collected both for rural and urban sectors. For ownership holdings, data on land owned, land leased in and out, types and terms of lease were collected. In regards to operational holdings, data on size, composition, tenurial form, land use pattern, extent of irrigation by source, fragmentation of holdings, drainage facilities and other related aspects of holdings were collected. Further, data on inventory of livestocks and agricultural machinery owned and otherwise possessed by the households were also collected.

In NSS, a plot of land is considered to be owned by a household if the right of permanent heritable possession with or without the right to transfer the title was vested on a member or members of the household. Land held in owner-like possession under long-term lease or assignment was also considered as land owned by the household. In a state where land legislation has provided for full proprietorship to erstwhile tenants, even though tenants did not pay full compensation, they were considered as having owner-like possession. Sometimes a tribal in accordance with traditional tribal rights from local chieftains or village/district council may possess a plot. Again, a tenant may occupy a plot while the right of ownership rested in the community. In both the cases, the tribal and the tenant are considered as owners because in all such cases the holders have the owner-like possession of land in question. A household ownership holding consists of all plots of land cultivated or not, owned by a household.

The land holding survey of 48th round was a repeat survey of the 26th round with some modification in item coverage and the method of data collection. While the reference period for collecting the data on operational holdings in the 26th round was the entire agricultural year (1970-71), in the 48th round information was collected separately for the kharif and rabi seasons of the agricultural year 1991-92 at the end of each season by making two visits to every sample household at an interval of nearly 6 months. In addition, particulars of operational holdings with the agricultural year 1991-92 as the reference period were also collected in the second visit i.e. rabi season. The 48th round covered all the main items of information on operational and ownership holdings as collected in the 26th round excluding the information on the use of fertilisers and pesticides and area under different crops.

In the NSS, an operational holding is defined as a techno-economic unit used wholly or partly for agricultural production and operated (directed/managed) by one person alone or with the assistance of others, without regard to title, size or location. The holding may consist of one or more parcels of land, provided they are located within the country and as a whole formed part of the same technical unit. A technical unit is a unit with more or less independent technical resources covering items like land, agricultural equipments and machinery, draught animals etc. Operational holdings also include all holdings used exclusively for livestock and poultry rising and for production of livestock and poultry products (Primary) and/or pisciculture whereas holdings put exclusively to uses other than agricultural production are not considered as operational holdings. Holdings operated by cooperative farms are also not considered as operational holdings.

The extended definition of agricultural production adopted in the 48th round includes:

1. Growing of field crops, fruits, grapes, nuts, seeds, tree nurseries (except those of forest trees), bulbs, vegetables and flowers both in the open and in glass houses.
2. Production of coffee, tea, cocoa, rubber, forest production in parcels of land which form part of the enumeration holding.
3. Production of livestock and livestock products, poultry and poultry products, fish, honey, rabbits, fur-bearing animals, and silkworm cocoons.

2.4.1.3 Analysis of Census Data

The distribution of operational holdings in Manipur is highly uneven. Table 2.11 shows that marginal holdings predominates the distribution of operational holdings over the decades. It accounts for nearly 50 percent of the total operational holdings but it occupied just over 21 percent of total area operated. Marginal and small holdings constitute as much as 83 percent of the total number of operational holdings occupying about 60 percent of area operated. While medium and large holdings constitute only about 2 percent, they account about 9 percent of total area operated. There has not been any remarkable change in the picture over the decades.

When the percentage distribution of operational holdings in the state is compared with all India level as shown in Table 2.12, there is a marked difference between the two. The marginal and small holdings predominate the

TABLE 2.11

Distribution of Operational Holdings in Manipur

Size Class	*Number*					*Area*				*(in hectare)*
	1970-71	*1976-77*	*1980-81*	*1985-86*	*1990-91*	*1970-71*	*1976-77*	*1980-81*	*1985-86*	*1990-91*
Marginal	32677 (40.88)	70186 (49.50)	65031 (47.77)	67314 (48.18)	68598 (48.29)	17452 (18.90)	38007 (24.02)	35583 (21.00)	36553 (21.05)	37820 (21.61)
Small	34270 (42.88)	49405 (34.84)	47605 (34.97)	48225 (34.51)	49062 (34.53)	40634 (44.00)	61026 (38.56)	65015 (38.38)	66211 (38.13)	67135 (38.37)
Semi Medium	11984 (14.99)	20060 (14.15)	20637 (15.16)	21093 (15.10)	21452 (15.10)	29213 (31.64)	48810 (30.85)	53408 (31.53)	54477 (31.38)	54892 (31.37)
Medium	974 (1.22)	2105 (1.49)	2832 (2.08)	3038 (2.17)	2916 (2.05)	4663 (5.05)	9926 (6.27)	15028 (8.87)	15523 (8.94)	14611 (8.35)
Large	22 (0.03)	34 (0.02)	30 (0.02)	55 (0.04)	43 (0.03)	378 (0.41)	478 (0.30)	373 (0.22)	873 (0.50)	523 (0.30)
All Size	79,927	1,41,790	1,36,135	1,39,725	1,42,071	92,340	1,58,247	1,69,407	1,73,637	1,74,981

Source: Government of Manipur: Agricultural Census of Manipur 1970-71, 1976-77, 1980-81, 1985-86 & 1990-91; Department of Agriculture.

Note: 1. The figures for 1970-71 agricultural censuses are not strictly comparable with those of the subsequent censuses because 1970-71 census relate to central district and 13 selected villages of south district only.

2. Figures in the parenthesis indicate the percentage distribution.

TABLE 2.12

Percentage Distribution of Operational Holdings (Manipur & All India)

Size Class	*1970-71*				*1990-91*			
	Number		*Area*		*Number*		*Area*	
	Manipur	*All India*	*Manipur*	*All India*	*Manipur*	*All India*	*Manipur*	*All India*
Marginal	40.88	50.98	18.90	8.98	48.29	59.44	21.61	15.04
Small	42.88	18.92	44.00	11.89	34.53	18.84	38.37	17.42
Semi Medium	14.99	15.04	31.64	18.50	15.10	13.06	31.37	23.19
Medium	1.22	11.17	5.05	29.75	2.05	7.11	8.35	27.04
Large	0.03	3.89	0.41	30.88	0.03	1.55	0.30	17.31
All Size	100	100	100	100	100	100	100	100

Source: 1. Government of Manipur: Agricultural Census of Manipur 1970-71, State Agricultural Census Office.
2. Government of Manipur: Report on Agricultural census 1990-91, Department of Agriculture.
3. Government of India: All India Report on Agricultural Census 1970-71, Department of Agriculture and Cooperation, Ministry of Agriculture.
4. Government of India: Agricultural Statistics At a Glance 2000, Directorate of Economics and Statistics, Department of Agriculture and Co-operation, Ministry of Agriculture.

distribution of operational holdings at all India level too, both in 1970-71 and 1990-91 censuses but the percentage of area operated by these two classes were much smaller than Manipur. In 1970-71, the two categories together constituted 83.76 percent and 69.90 percent of total operational holdings operating on 62.90 percent and 20.87 percent of the total area operated in Manipur and all India level respectively. Similarly in 1990-91, the two classes together accounted for 82.82 percent and 78.28 percent of the total operational holdings operating on 59.98 percent and 32.46 percent of the total operated area in Manipur and all India respectively. Besides, in 1970-71, small operational holdings constitute 42.88 percent of the total operational holdings accounting for 44.00 percent of the total operated area of Manipur while at all India level the corresponding figures are 18.92 percent and 11.89 percent respectively. In 1990-91, the corresponding figures are 34.53 percent and 38.37 percent respectively in Manipur while the all India figures are 18.84 percent and 17.42 percent respectively. On the other hand, large holdings in the state constitute only 0.03 percent of the total number of operational holdings, which operate less than 0.5 percent of the total operated area, both in 1970-71 and 1990-91 censuses. However, at all India level the corresponding figures are 3.89 percent and 30.88 percent respectively in 1970-71 and 1.55 percent and 17.31 percent respectively in 1990-91.

The average size of operational holdings for different size classes in Manipur, as shown in Table 2.13, fluctuates from census to census. However, largest variation can be observed in the large operational holdings when it decreased by 27.65 percent from 17.18 hectares in 1970-71 to 12.43 hectares in 1980-81. It increased to 15.87 hectares in 1985-86 but again fell to 12.16 hectares in 1990-91. In the case of marginal holdings, the average size remained relatively stable through out the decades while that of small holdings, it become stable at 1.37 hectares from 1980-81 onwards.

TABLE 2.13

Average Size of Operational Holdings in Manipur

(in hectare)

Size Class	*1970-71*	*1976-77*	*1980-81*	*1985-86*	*1990-91*
Marginal	0.53	0.54	0.55	0.54	0.55
Small	1.19	1.24	1.37	1.37	1.37
Semi-Medium	2.44	2.43	2.59	2.58	2.56
Medium	4.79	4.72	5.31	5.11	5.01
Large	17.18	14.06	12.43	15.87	12.16
All Holdings	1.16	1.12	1.24	1.24	1.23

Source: Same with Table 2.11

TABLE 2.14

Average Size of Operational Holdings for Manipur and All India in 1970-71 and 1990-91 Agricultural Census

(in hectare)

	1970-71		*1990-91*	
Size Class	*Manipur*	*All India*	*Manipur*	*All India*
Marginal	0.53	0.40	0.55	0.39
Small	1.19	1.44	1.37	1.43
Semi-Medium	2.44	2.81	2.56	2.76
Medium	4.79	6.08	5.01	5.90
Large	17.18	18.10	12.16	17.33
All Holdings	1.16	2.28	1.23	1.55

Source: Same as Table 2.12.

When the average size of operational holdings of different size class in the state is compared with all India average both in 1970-71 and 1990-91, only the average size of marginal holdings in Manipur is bigger than all Indian average. Whereas, in all other categories, all India average is larger than that of Manipur. However, the average size of different size class at all India level has declined over the decades which are contrary to the case of Manipur. This shows that marginalisation of operational holdings at all Indian level is more severe than in Manipur. Further, the distribution of operational holdings in the state is much more equal than the all India level. This inference is further endorsed by the Gini Coefficients of the operational holdings shown in Table 2.15.

TABLE 2.15

Gini Coefficients of Operational Holdings Based on Agricultural Censuses

State / Country	*1970-71*	*1976-77*	*1980-81*	*1985-86*	*1990-91*
Manipur	0.3116	0.3295	0.3507	0.3536	0.3456
All India	0.6202	0.6141	0.6038	0.5925	0.5794

Source: Same with Table 2.12.

The above Table shows that inequality in the distribution of operational holdings is much higher at all India level. However over the decades the inequality in the distribution of operational holdings has been increasing in Manipur even though there is a marginal fall in 1990-91, while for all India it has been declining continuously over the corresponding period.

A census based comparative study on the nature of change in the structure of operational holdings over 1985-86 to 1990-91 shows that the all India experience can be extended to Manipur up to some extent. The difference is only the pace of change. It can be confirmed from Table

TABLE 2.16

Percentage Changes in Number of Holdings and Operated Area According to Farm Size Groups in 1990-91over 1970-71 and 1985-86 in Manipur and All India

Size Class	Over 1970-71				Over 1985-86			
	Manipur		All India		Manipur		All India	
	Number	Area	Number	Area	Number	Area	Number	Area
Marginal	109.93	116.71	75.11	70.99	1.91	3.47	12.90	12.94
Small	43.16	65.22	49.54	49.50	1.74	1.40	12.11	12.13
Semi-Medium	79.01	87.90	30.35	27.92	1.70	0.76	5.06	4.66
Medium	199.38	213.34	-4.44	-7.22	-4.02	-5.88	-4.24	-5.07
Large	95.45	38.36	-40.20	-42.76	-21.82	-40.09	-13.76	-13.16
All Size	77.75	89.50	50.17	2.08	1.68	0.77	9.76	0.57

Source: Same as Table 2.12.

2.16. Firstly, in Manipur the number and area under marginal operational holdings have increased only by 1.91 percent and 3.47 percent respectively. But at all India level the corresponding figures were 12.9 percent and 12.94 percent respectively. A similar picture can be observed in the case of small and semi-medium holdings as well.

Secondly, the decrease in the number and area of large operational holdings is much slower at all India level. However, in the case of medium holdings the pace of variation is almost similar both in Manipur and all India level. Further, it has been justified that large chunk of operated area at all India level continue to be operated by few people. However, it is fast disappearing in Manipur.

Finally, the increase in the number of operational holdings at all India level is much faster than Manipur whereas the increase in area operated is slower. This may be because of faster increase in number of marginal and small operational holdings with only marginal rise in area operated.

Operated area need not be identical with owned area. Land owned has been defined as a piece of land for which the cultivator has the right of permanent heritable possession. This would include land held from government or others under grant or lease or assignment with right of permanent heritable possession with or without right of transfer and land operated under perpetual lease.[22] Agricultural census data show, however, that owned area can be approximated by operated area.

As per agricultural census of Manipur 1970-71, area owned and self-cultivated constituted 86.07 percent of total cultivated areas and 90.12 percent of the total operational holdings. The percentage of area under lease was only

TABLE 2.17

Percentage of Partly/Wholly Rented Operational Holdings in Manipur

	1970-71		*1990-91*	
Size Class (ha.)	*Partly Rented*	*Wholly Rented*	*Partly Rented*	*Wholly Rented*
Below 0.5	0.42	0.53	0.62	11.17
0.5 - 1.0	2.73	1.99	1.09	5.59
1.0 - 2.0	7.76	5.00	1.69	4.61
2.0 - 3.0	11.10	7.17	2.70	2.89
3.0 - 4.0	12.00	4.71	2.21	2.57
4.0 - 5.0	16.57	3.50	4.42	3.40
50 - 10.0	9.49	3.16	6.24	1.04
10 - Above	-	-	-	-
All Size	6.04	3.84	1.52	5.77

Source: 1. Government of Manipur: "Agricultural Census of Manipur 1970-71", State Agricultural Census Office.

2. Government of Manipur: "Report on Agricultural Census 1990-91", Department of Agriculture.

8.79% of the total cultivated areas of the state. Fixed produce terms of leasing is the most popular with 99.68 percent of the total area under lease. In 1990-91 census, area owned and self-cultivated increased to 93.8 percent of the total cultivated area and 92.48 percent of holdings were wholly owned and self operated. The percentage of area on lease either fully or partly has came down to 4.91 percent of the total cultivated area of the state. Also, the terms of leasing have become more diversified. The percentage of area leased on fixed produce was only 44.52 percent of the total area on lease either fully or partly. The shares of produce and usufructuary mortgage have become more popular with 23.58 percent and 21.28 percent respectively. Fixed money term is the least practised with only 10.62 percent.

2.4.1.4 Analysis NSS Data

The importance of marginal and small size operational holdings can be confirmed from the NSS data as well. According to the 48th round, the two categories constitute 92.15 percent of the total number of operational holdings accounting for 74.12 percent of the total area operated. However, unlike the census data, the number of marginal operational holdings has increased at a much faster pace without an equivalent rise in the area operated. The medium and large operational holdings in the 48th round have almost disappeared. Due to the increase in the number of

TABLE 2.18

Distribution of Operational Holdings in 26th, 37th & 48th Round of NSS

	Number of Operational Holdings ('00)			*Area of Operational Holdings('00 hect.)*		
Size Class	*26th Round*	*37th Round*	*48th Round*	*26th Round*	*37th Round*	*48th Round*
Marginal	595 (51.56)	839 (52.67)	1523 (72.49)	324 (24.96)	390 (25.61)	579 (37.94)
Small	410 (35.53)	611 (38.36)	413 (19.66)	565 (43.53)	731 (47.99)	552 (36.18)
Semi-Medium	135 (11.70)	130 (8.16)	163 (7.76)	336 (25.88)	307 (20.16)	389 (25.49)
Medium	13 (1.12)	12 (0.75)	2 (0.09)	65 (5.01)	57 (3.74)	6 (0.39)
Large	1 (0.09)	1 (0.06)	--	8 (0.62)	38 (2.50)	--
All Sizes	1154	1593	2101	1298	1523	1526

Source: 1. NSS, 26th Round, Report No. 215.
2. NSS, 37th Round, Report No. 331.
3. NSS, 48th Round, Report No. 407.

Note: Figures in the parenthesis indicate the percentage.

TABLE 2.19

Percentage Distribution of Operational Holdings in Manipur and All India in 26th and 48th Round of NSS

Size Class	*26th Round*				*48th Round*			
	No. of household		*Area operated*		*No. of household*		*Area operated*	
	Manipur	*All India*	*Manipur*	*All India*	*Manipur*	*All India*	*Manipur*	*All India*
Marginal	51.56	46.23	24.96	9.10	72.49	65.15	37.94	15.59
Small	35.53	22.10	43.53	14.65	19.66	16.60	36.18	18.42
Semi-Medium	11.70	17.54	25.88	22.42	7.76	11.20	25.49	23.81
Medium	1.12	11.04	5.01	30.40	0.09	5.77	0.39	26.30
Large	0.09	3.09	0.62	23.32	-	1.28	-	15.88
All Size	100	100	100	100	100	100	100	100

Source: 1. NSS 48th Round, Report No. 407.

2. Laxminarayan, H. and S.S. Tyagi (1982): "Changes in Agrarian Structure in India", Agricole Publishing Academy, New Delhi, p. 46.

marginal holdings, the total number of operational holdings has increased by about 32 percent between 37th and 48th rounds. However, the total area operated almost remained stagnant.

When the percentage distribution of operational holdings in the state is compared with all India level, the two revealed quite a contrasting picture as shown in Table 2.19.

Table 2.19 shows that marginal and small holding categories constitute a large proportion of the number of operational holdings both in Manipur and all India level. However, the share of area operated by these two categories of holders in Manipur and all India level differ significantly. While in Manipur the two categories operate over 74.12 percent of the total area operated in the state, it is only 34.01 percent at all India level. The percentage of medium and large holdings constitute only about 0.09 percent of the total operational holdings and they account for only 0.39 percent of the total operated area in Manipur. At all India level the corresponding figures are 7 percent and 42.18 percent respectively. All this implies that the operational holdings are much more equally distributed in Manipur than at all India level.

TABLE 2.20

Average Size of Operational Holdings in Manipur

(in hectares)

Size Class	*26th Round*	*37th Round*	*48th Round*
Marginal	0.54	0.46	0.38
Small	1.38	1.20	1.34
Semi-Medium	2.49	2.36	2.39
Medium	5.00	4.75	3.00
Large	8	38	--
All Size	1.12	0.96	0.73

Source: Same with Table 2.18.

When we look at the average size of operational holdings in different size class, as shown in Table 2.20, in Manipur the average size of overall operational holdings has been continuously declining from 1.12 hectare in 26th round to 0.96 hectare in 37th round and then to 0.73 hectare in 48th round. Among the different size classes, marginal and medium holdings also show the same declining trend. Similarly, the average size of small and semi-medium holdings, though fluctuate marginally from round to round, shows downward trend. This indicates the increasing marginalisation of operational holdings in Manipur. Let us now see the comparative picture of average size of operational holdings in Manipur and all India, which is shown in Table 2.21.

TABLE 2.21

Average Size of Operational Holdings in Manipur and All India in 26th and 48th Round of NSS

	Manipur		*All India*	
Size Class	*26th Round*	*48th Round*	*26th Round*	*48th Round*
Marginal	0.54	0.38	0.44	0.30
Small	1.37	1.34	1.46	1.41
Semi Medium	2.49	2.39	2.81	2.69
Medium	5.00	3.00	6.04	5.78
Large	8.00	--	16.56	15.68
All Sizes	1.12	0.73	2.19	1.27

Source: NSS 26th Round, Report No. 215 and 48th Round, Report No. 407.

When average size of operational holdings in Manipur is compared with all India, it revealed that in both 26th and 48th round of NSS average size of marginal holdings in Manipur is slightly larger than that of all India average while in all other remaining classes all India average is much larger than that of Manipur. It becomes more prominent in the case of large holdings. At all India level the gap between the average size of marginal and other holdings goes on increasing with the increase in the size of operational holdings. Similarly, the overall average size of operational holdings at all India level is much larger than that of Manipur. Thus, we can conclude that the marginalisation of operational holdings is more severe at all India level than in Manipur. It can be further endorsed by the difference in value of Gini coefficient of inequality as shown in Table 2.22.

TABLE 2.22

Gini Coefficients for Operational Holding of Manipur and All India

	26th Round	*48th Round*
Manipur	0.3252	0.3681
All India	0.5710	0.6078

Source: Same as Table 2.19.

The value of Gini coefficient of inequality in Manipur keeps on increases from 0.3252 in 26th round to 0.3681 in 48th round. This implies an increasing marginalisation of operational holdings in Manipur. However, in both the rounds its value at all India level is much larger than that of Manipur. Thus, we can conclude that the distribution of operational holdings in Manipur is more equal than all India level.

A comparison on the percentage change in operational holdings of the later two rounds of NSS i.e. 37th and 48th on operational holdings in Manipur shown in Table 2.23 revealed that large size holdings have completely vanished in the 48th round. While the number and area operated by small and Medium holdings have declined there is a steep

TABLE 2.23

Percentage Change in Operational Holdings in 48th Round Over 37th Round of NSS in Manipur

Size Class	*No.of operational holdings (in '00)*			*Area of operational holdings (in '00 hect.)*		
	37th	*48th*	*%*	*37th*	*48th*	*%*
Marginal	839	1523	81.53	390	579	48.46
Small	611	413	-32.41	731	552	-24.49
Semi-Medium	130	163	25.38	307	389	26.71
Medium	12	2	-83.33	57	6	-89.47
Large	1	0	-100	38	0	-100
All Sizes	1593	2101	31.89	1523	1526	0.20

Source: Same with Table 2.18.

rise in the number of marginal holdings. The area operated by them has increased by about 48 percent. On the other hand, the total number of operational holdings has increased by 31.89 percent while the area operated increased by just 0.2 percent. This shows that while the number of population dependent on agriculture continued to increase, the scope for the extension of agricultural land seems to have reached the saturation level. As a result the marginalisation of operational holdings is getting progressively steeper over the decade.

2.4.1.4 NSS Vs Agricultural Census Results[24]

The first agricultural census and the 26th round of NSS have the same reference period. However in the case of 37th and 48th round the reference period was major crop seasons of the agricultural year 1981-82 and 1991-92 respectively. While on the corresponding census i.e. 3rd and 5th the reference period was the agricultural year 1980-81 and 1990-91 respectively. Since the reference periods were close to each other and were more or less normal agricultural years, the results are comparable.

As distinct from the household enquiry method of the NSS, the agricultural census data on operational holdings were compiled from the land records in the areas where such data are available on complete enumeration basis. In non-land record areas like Manipur the relevant data were collected by interviewing the sample households.

In the land holding surveys of NSS, the required information was collected by direct inquiries of households of the sample, and hence covers only the household sector. In agricultural census, besides the operational holdings belonging to the households the holdings of the non-household sector such as tea garden, coffee plantations managed by companies, institutions, state agricultural farms, cooperatives farms etc. were also covered. This exerted a downward pull to the NSS estimations.

The agricultural census delimited an operational holding within a tehsil/village, while the NSS extended the geographical boundary to the whole state in 26th round and whole of India in 37th and 48th rounds provided the parcels formed part of the same techno-economic unit. In the agricultural census, the agricultural holdings were listed in the village of residence of the respective cultivators.

In the NSS surveys, production of livestocks and poultry and its products, fish, honey, rabbits, fur bearing animals and silkworm cocoon were included. However these items were not covered in the agricultural census.

Similarly in the NSS survey, the total area covered in an operational holding includes residential area under farm houses located outside the cultivated area. But it is excluded in the agricultural census.

In the agricultural census, exclusively current fallow holdings were taken as operational holdings. They are excluded in the NSS.

TABLE 2.24

Size Distribution of Operational Holdings NSS Vs. Census

	No. of Operational Holdings ('00)			*Area Operated in ('00 Ha.)*		
Size Class	*1990-91 Agri. Census*	*48th Round*	*Percentage Difference*	*1990-91 Agri. Census*	*48th Round*	*Percentage Difference*
Marginal	686 (48.28)	1523 (72.49)	122.01	378 (21.61)	579 (37.194)	53.17
Small	491 (34.55)	413 (19.66)	-15.89	671 (38.37)	5512 (36.18)	-17.73
Semi-Medium	215 (15.13)	163 (7.76)	-24.17	549 (31.37)	389 (25.49)	-29.14
Medium	29 (2.04)	2 (0.09)	-93.10	146 (8.35)	6 (0.39)	-95.89
Large	*	--	--	9 (0.50)	--	-100
Total	1421	2101	47.85	1753	1526	-12.95

Source: 1. Government of Manipur: Report on Agricultural Census 1990-91, Department of Agriculture.

2. NSS, 48th Round, Report No. 407.

Note: * Less than 100

1. Figures in parenthesis indicate percentage of column total.

2. $\text{Percentage Difference} = \frac{\text{NSS figure - Census figure}}{\text{Census figure}} \times 100$

A meaningful comparison between the agricultural census 1970-71 and NSS 26th round data on the distribution of operational holdings over different size classes is not possible because of limited coverage of the former.

However, the subsequent two rounds of NSS viz., 37th land 48th round can be compared with corresponding census of 1980-81 and 1990-91 in a meaningful way. Singh, E. B. K. (1992) discussed the result of the comparison between NSS 37th round with 1980-81 census. Therefore, in the present analysis we shall concentrate on the comparison between 48th round of NSS with 1990-91 agricultural census. It reveals that NSS figure exceeds the census figure in the case of marginal holdings while the reverse is true in the remaining size categories viz., Small, Semi-medium, Medium, and Large size holdings. The difference in the number of marginal operational holdings is far greater than the area operated by them. The result of the comparison is almost similar to that of NSS 37th round with 1980-81 census except in the case of small holdings where the NSS figure is larger than the census figure. The two categories still constitute over 91 percent of total number of operational holdings and around 74 percent of operational area in the 48th round of NSS as well. This is because of increase in the number and area operated under marginal holdings and decrease in the number and area operated by small holdings in almost similar magnitude.

On the other hand, smaller number of small, semi-medium and large size holdings in the NSS may be the result of larger dispersal of operational holdings in these categories. In the NSS the boundary of operational holdings could extend over the whole of India provided the parcels formed part of the same techno-economic unit whereas in the agricultural census it was the village residence of the cultivator. The implication of this difference in the definition of operational holding is that parcels spread over different villages will be counted as different operational holdings in the census and as single operational holding in the NSS provided the parcels formed part of the same techno-economic unit. The larger area under these operational holdings in the census may be attributed to inclusion of current fallow land in the estimation of area under operational holdings. It implies that substantial part of land under these categories was rendered as current fallow.

The operational holdings need not be distributed compactly in a single block. It may consist of parcels/ fragments within a village/tensil in an agricultural census and over the entire state in NSS.

By definition an operational holding includes all those fragments dispersed widely. The dispersal of holdings affects the management and utilization of land resources effectively.

The degree of dispersal of operational holdings will be examined on the basis of the extent of parcelisation of holdings. A parcel has been defined as a piece of land entirely surrounded by other operational holding. It might consist of one or more plots. A holder having parcels within and outside the village of his residence will face manifold problems both in terms of material and manpower management.

Table 2.25 shows the extent of parcelisation in the state over the decades. The average number of parcels in marginal class has declined while the average area per parcel has increased continuously since 1976-77. In the remaining classes including all size classes there is a wide fluctuation in the figures from one census to another census. But when we compare the result of two censuses i.e. 1991-92 over 1986-87 census, the extent of parcelisation of operational holding in the state has declined tremendously in all size classes. As a result the average size of a parcel has also increased. Thus, we can conclude that since the

TABLE 2.25

Parcelisation in the State between 1970-71 to 1991-92

Size Class	*Average number of Parels per Operational Holding*					*Average area per Parcel*				
	1970-71	*1976-77*	*1981-82*	*1986-87*	*1991-92*	*1970-71*	*1976-77*	*1981-82*	*1986-87*	*1991-92*
Marginal	1.04	1.08	1.07	1.01	1.00	0.52	0.50	0.51	0.54	0.55
Small	1.70	1.68	2.02	1.71	1.04	0.70	0.74	0.65	0.80	1.32
Semi-Medium	2.48	3.70	3.03	3.11	1.06	0.98	0.66	0.91	0.83	2.41
Medium	3.73	4.21	4.48	6.45	1.13	1.29	1.12	1.24	0.79	4.50
Large	11.36	6.09	7.00	17.40	1.60	1.65	2.31	1.79	0.76	8.88
All Size	1.57	1.71	1.77	1.69	1.03	0.73	0.65	0.71	0.73	1.20

Source: 1. Government of Manipur: Agricultural Census of Manipur 1970-71, State Agricultural Census Office.

2. Sarao, B.S. (1987): All India Report on Input Survey 1976-71, Vol.1, Department of Agriculture and Rural Development, Ministry of Agriculture and Rural Development, P. 15.

3. Government of Manipur: Agricultural Census 1980-81 and Input Survey 1981-82 of Manipur, Department of Agriculture.

4. Government of Manipur: Input Survey 1986-87 & 1991-92, Department of Agriculture, Ministry of Agriculture.

later part of 80s the parcelisations in the state have declined as farmers seems to realise the defect of fragmented holdings.

2.4.2 Distribution of Assets

Assets constitute a very important factor in the field of production process and agricultural production is no exception. The productivity of agriculture depends on the types of assets in use. The productive assets raise the productivity and thus generate income for the asset owners who either use them or supply them to others on rent. Information on inventory of assets has been collected in a series of decennial survey conducted by the Reserve Bank of India (RBI) and National Sample Survey Organization (NSSO). So far five surveys have been conducted in this field. The first two surveys on distribution of assets were conducted by RBI in All India Rural Credit Survey, 1951-52 and All India Rural Debt and Investment Survey, 1961-62. However, from the third survey onwards responsibility of conducting the survey was given to the NSSO. This organization undertook the All India Debt and Investment Survey in its 26th round (July 1971- September 1972). The fourth decennial survey on Debt and Investment was conducted in the NSS 37th round (January-December 1982) and the fifth survey was conducted in the NSS 48th round (January-December, 1992)

Household assets were defined to include all items that were owned by the household which had money value such as land, building, livestock, agricultural machinery and implements, non-farm business equipment, all transport equipment, durable household goods, advances in cash and in kind, shares in companies and cooperative societies, banks etc., units of various mutual funds, national saving certificates and the like, contribution made towards provident fund, payment of insurance premium, deposits in companies, banks, post offices, and with individuals. The earlier AIDIS did not include crops standing in the fields and stock of commodities held by the household in the household assets. Currency notes and coins in hand were also not considered as assets in any of the earlier surveys owing to difficulties in collecting reliable data on them. However, in the 48th round of NSS, an attempt was made to collect information on the amount of cash in hand of the household as on the date of survey.[25]

All rural households operating at least 0.002 hectare of land during the agricultural year 1991-92 was treated as cultivator household. Cultivator households constituted 75.99 percent of rural households in Manipur in the 48th round of NSS. In the following section, we shall examine the asset structure of rural cultivator households. The study on asset structure of cultivator households is imperative because of their importance in rural structure in terms of number and their relation with agricultural sector. The comparison will be made with some selected states on the basis of their performance of growth of cereal output.

Sawant (1997) classified the 17 major states of India into five categories according to the levels of growth rate in cereal and pulse production. States with high growth rate (above 4%) during 1981-95 were West Bengal and Haryana. States with moderate growth rate (2.1% to 4%) were Orissa, Punjab, Assam, Bihar, H.P., Rajasthan, U.P., Karnatraka, M.P. and Tamil Nadu. Low growth states (up to 2%) were J. K., Andhra Pradesh and Maharashtra. Gujarat and Kerala were categorised as zero or highly unstable growth rate and negative growth rate respectively.[26] However in our present study we shall classify these 17 states into three categories only viz. high (above 4%), moderate (2.1% to 4%) and low (up to 2%) growth rate. States with negative and zero or highly unstable growth states i.e. Kerala, Gujarat were clubbed together with low (up to 2%) growth states.

TABLE 2.26

Percentage Share of Assets of Rural Cultivator Households (48th Round, NSS)

Assets	*Punjab*	*Haryana*	*Maharashtra*	*Assam*	*Manipur*	*All India*
Land	77.84	80.80	69.39	60.61	55.87	67.61
Building	11.80	9.78	15.95	21.28	27.18	19.13
Livestocks and Poultry	1.91	2.42	4.04	6.71	3.18	3.45
Machinery and equipment	5.82	4.56	4.69	1.82	4.33	3.80
Household durables	2.17	2.16	4.56	7.40	7.96	5.15
Financial assets	0.44	0.28	1.23	2.11	1.46	0.81
Loan receivable	0.02	--	0.14	0.07	0.02	0.05

Source: NSSO (1998): "Household Assets and Liabilities as on 30-6-1991", Debt and Investment Survey, 48th Round, Report No. 419.

Table 2.26 gives the comparative picture of shares of major asset categories in the states selected from all the three categories classified above. Haryana belonged to the category of high growth state, Punjab and Assam belong to the category of moderate growth states and Maharashtra belonged to the low growth category. As per the growth rate of rice production in Manipur during 1972-1999, Manipur belongs to the low growth category. The comparative study of the share of assets reveals that the share of land in Manipur is substantially lower than all the states and more particularly high growth state Haryana. In the case of building and household durables, Manipur has the highest percentage share but it is comparable with Assam to some extent. The share of livestocks and poultry in the state is comparable with Haryana but substantially lower than Assam which is a moderate growth state. The share of machinery and equipments in the state is comparable with Punjab, Haryana and Maharashtra but its share is quite low in Assam. In the states where mechanization is gaining momentum the share of livestocks declines.

Apart from the difference in the share of different assets categories, the average value of items per reporting household in Manipur is lower than the states with high rate of growth. This indicates absolute inadequacy in the asset base of cultivator households in Manipur.

TABLE 2.27

Average Value of the Item Per Reporting Cultivating Household in Rs. as on 30-6-1991 (48th Round, NSS)

Assets	*Punjab*	*Haryana*	*Maharashtra*	*Assam*	*Manipur*	*All India*
Land	478701	440065	92933	44251	52776	96216
Building	72576	53248	21363	15538	25680	27231
Livestock and Poultry	11763	13205	5408	4897	3000	4909
Machinery and equipment	35783	24828	6278	1329	4094	5404
Household durables	13334	11755	6104	5402	7516	7325
Financial assets	2695	1507	1641	1542	1377	1158
Loan receivable	113	18	194	49	20	65
Total	*614965*	*544626*	*133921*	*73008*	*94463*	*142308*

Source: Same as above.

The average value of land per reporting cultivator household in Manipur is only 11.99 percent of the value in Haryana. Similarly, the value of machinery and equipments is 16.49 percent of the value of in Haryana. In short, value of all the asset categories in Manipur is much lower than the value in states with high growth rate in cereal production.

Unfortunately the gap is getting wider than the NSS 37th round. However, when compared with all India level, except for livestocks and poultry, the difference in the value of all other asset categories is progressively declining. In the case of household durables and financial assets, value in Manipur is more than all India level in the 48th round. Even though Manipur has the highest average value of assets per reporting cultivator household among the North Eastern states of India viz. Assam, Tripura, Mizoram, Nagaland, Meghalaya and Arunachal Pradesh in the 48th round, it is exceedingly low when compared with Haryana and all India. It is only 17.34 percent of the value of Haryana and 66.38 percent of the value of for all India.

TABLE 2.28

Percentage Share of Assets of the Rural Cultivator Households in Manipur and All India (Different Rounds of NSS)

	26th Round		*37th Round*		*48th Round*	
Assets	*Manipur*	*All India*	*Manipur*	*All India*	*Manipur*	*All India*
Land	53.31	69.0	57.43	64.1	55.87	67.6
Building	24.58	16.5	21.96	19.5	27.18	19.1
Livestocks and poultry	5.81	6.5	7.68	5.0	3.18	3.4
Machinery and equipment	2.44	2.7	2.91	3.7	4.33	3.8
Household durables	5.57	4.2	8.99	6.6	7.96	5.2
Financial assets	--	0.8	--	1.0	1.46	0.8
Loan recievable	--	0.6	--	0.1	0.02	0.1

Source: 1. NSSO (1998): "Household Assets and Liabilities as on 30-6-1991", Debt and Investment Survey, 48th Round, Report No. 419.
2. Singh, E. B. K. (1992): Op. Cit., P.103.

In examining the changes in the shares of assets over time it should be noted that the category of vacant house sites in the 26th round has disappeared from the 37th round onwards, instead the new category of non-farm business equipments appears.

In the rural cultivator household sector the percentage of the value of land to the value of total assets was the highest among the major asset groups. However the share of land in Manipur has fallen from 57.43 percent in the 37th round to 55.87 percent in the 48th round. In the case of all India the share has increased from 64.1 percent to 67.61 percent over the corresponding rounds. In the case of building the reverse is true. While the shares of livestocks and poultry and household durables have declined from the previous round both in Manipur and all India, the share of machinery and equipments has increased and more particularly in Manipur.

The pattern of changes in the shares of different assets can be interpreted in the following ways:

1. The increase in the share of machinery and equipments is an encouraging sign as it is a characteristic of high growth states. It indicates an increasing mechanization of agriculture and development of non-farm business. In other words, capital-intensive method of cultivation has been increasingly adopted in the state.
2. The decline in the share of land and increase in the share of buildings and almost stagnant share of household durable are a major source of concern. These two asset categories viz., buildings and household durables together constitute 35.14 percent of the total assets per reporting rural cultivator household. It is a characteristic not shared with high growth states.

These results corroborate the finding of Singh, E. B. K. (1992) on the study on asset distribution in the state based on two earlier rounds of NSS. The high and still increasing share of buildings and stagnant share of household durable assets may be an indication of better position of housing in the rural areas of Manipur. It may also indicate the minimum needs orientation of the cultivators in Manipur unlike the profit orientation of the cultivators in agriculturally prosperous states like Punjab. The late beginning of the process of commercialisation and late exposure to consumerism may have prompted the farmers to become more minimum needs oriented. It is pertinent to point out that almost every dynamic section of the society exhibits a tendency to go for consumption goods rather than strengthening the source of dynamism.[27]

TABLE 2.29

Gini Coefficients of the Distribution of Assets in 1991 for Rural Cultivator Households (NSS 48th Round)

Assets	*Punjab*	*Haryana*	*Maharashtra*	*Assam*	*Manipur*	*All India*
Land	0.3042	0.2936	0.5335	0.4731	0.5082	0.5994
Building	0.1747	0.1662	0.3630	0.4291	0.4011	0.4418
Livestocks and poultry	0.1735	0.1855	0.3169	0.2339	0.1303	0.3314
Machinery and equipments	0.3110	0.3063	0.6483	0.4137	0.7309	0.6598
Household durables	0.1406	0.2077	0.3633	0.3817	0.3439	0.4505
Financial assets	0.0899	0.1746	0.6302	0.1769	0.5928	0.5787
All assets	0.2824	0.2769	0.4960	0.4350	0.4566	0.5544

Source: NSSO (1998): "Household Assets and Liabilities as on 30.6.1991", Debt and Investment Survey, 48th Round, Report No. 419.

Table 2.29 shows the degree of inequality as measured by Gini coefficient. It reveals that there seems to be an inverse relationship between the growth of cereal output and inequality in the distribution of asset. High growth states like Haryana are characterized by low degree of inequality in distribution of assets. Similarly, in the case of Punjab, a state with moderate growth rate of cereal production during 1981-1995, the distribution is found to be more equal both item-wise and overall than Manipur. In the lower growth states the extent of inequality in the distribution of assets is higher than high growth states. When we compare the inequality in distribution with all India level, except for machinery & equipments and financial assets, in all other asset categories as well as over all assets the distribution is more equal in Manipur than all India. High degree of inequality in machinery and equipment and low degree of inequality in livestocks & poultry in Manipur indicates majority of cultivator households in rural sector continue to depend heavily on bovine source of power. It is only the rich and affluent farmers who own and use the agricultural machinery & equipments. Even though the share of machinery & equipments is comparable with high growth states like Haryana, mechanisation which has direct bearing with capital formation is far from satisfactory. The distribution of assets among rural cultivator can no longer be regarded as static in nature. It seems to vary with the growth in agricultural productivity.

Commenting on the question of equality in the distribution of assets in the context of structural change Bhalla (1987) writes:

> "It is useful to distinguish, first, that inequality in asset distribution which stands in the way of increasing the productive capacity of scarce land and abundant labour. Where that is the situation most economists would find in it a persuasive

> case for asset redistribution. . . . . -But suppose that the observed concentration of assets is, on the contrary, an integral part of a process which is increasing employment and labour productivity and making possible improvements in the standard of living of workers who have no such assets at all? Then what?"[28]

She further comments that capital accumulation and its concentration are characteristic features of technologies which generate economies of scale and which increase labour productivity and the real income of the working class.

2.5 Conclusion

In this chapter we have examined the structure of agrarian economy of Manipur with special reference to the distribution of land and non-land assets. The following findings emerge;

1. The inequality in the distribution of operational holdings has been increasing in Manipur while for all India it has been decreasing despite the fact that operational holdings are distributed much more equally in Manipur than in all India.
2. The census-based results on the pattern of change of marginal operational holdings do not agree with that of NSS. According to the census report, the marginal operational holding in Manipur remained more or less in same proportion. However, the NSS report showed an increasing trend. This discrepancy may be attributed to difference in the scope of agricultural production covered. Agricultural production in the NSS covers much more activities than in the census. Wider scope and the rising

trend in the number of area operated by marginal holdings may be interpreted as a growing diversification of the activity matrix of the rural households.

3. According to 1990-91 census, parcelisation in the state in all size classes including overall size has declined tremendously from 1985-86 census. It shows that the farmers are becoming increasingly conscious about the defects in the operation of fragmented holdings.
4. The asset structure of the cultivators in rural Manipur is quite different from states with high growth rate in cereal production. The averages values of assets-overall and item-wise are lower than high growth states. Similar results can be witnessed when compared with all India except for household durables and financial assets whose average values are higher in Manipur than in all India. Only the share of machinery & equipments is somewhat comparable with high growth states. While the share of land is substantially lower, the shares of building and household durable assets are substantially larger than those of high growth states. This may be attributed to the minimum needs orientation of cultivators in Manipur arising out of late beginning of commercialisation of agriculture and rapid exposure to consumerism.

Thus, what emerges after the detailed study of the agrarian structures of Manipur is an agriculture dominated by marginal and small farmers characterized by low degree of inequality and low degree of mechanization. It underscores the scope for raising output growth by enhancing the degree of mechanisation. The low degree of inequality indicates the possibility of sharing equitably the benefits of further agricultural development.

NOTES

1. Government of Manipur (1999): *Statistical Handbook of Manipur*, Directorate of Economics and Statistics.
2. ICAR (1992): *Soils of Manipur for Land Use Planning*, National Bureau of Soil Survey and Land Use Planning, Nagpur.
3. ICAR (1986): *Annual Report*, ICAR Research Complex for NEH Region, Manipur Centre, Babupara .
4. NCAER (1961): *Techno-Economic Survey of Manipur*, p. 12.
5. GOI (2001): Agricultural Development in Eastern and North Eastern India, TFYP Working Group Report Sr. No. 40/ 2001, Planning Commission, p .8.
6. Nagaland Environment Protection and Economic Development (NEPED) were initiated in 1994 in collaboration with Canadian Government. It has pioneered an excellent method of upgrading jhum by interposing components of agro-forestry through assisted tree planting of selected fast growing economic timber, the menu being a product of a meticulous exercise in biodiversity mapping, documentation and breeding of plant material for widespread propagation, *ibid.*, pp. 8-9.
7. Bio-village is an integrated approach with an aim to achieve a sustainable and equitable utilization of natural resources and bio-diversity and for linking conservation and commercialization in a mutually reinforcing manner. It aims to provide opportunities for production and marketing of organic products, health foods, eco- and bio-goods, herbal medicines etc. for the growing market for these commodities in India and abroad, *ibid.*, pp. 67-68.
8. *ibid.*, p. 68.
9. A good deal of literature on growth models is available in Singh, E.B.K. (1992): *op. cit.*
10. Minhas, B.S. and A. Vaidyanathan (1965): "Growth of Crop Output in India, 1951-1954 to 1958 - 1961: An Analysis of Component Elements" *Journal of the Indian Society of Agricultural Statistics*, vol. XVII, no. 2.

11. Dharam Narain (1976): *Growth of Productivity in Indian Agriculture*, occasional paper no. 93, Cornell University.
12. Desai, Bhupat M. and N.V. Nambooderi (1977): "Strategy and Sources of Growth in Crop-agriculture" in Bhupat M. Desai (eds.): *Agricultural Development Paradigm for the Ninth plan under New Economic Environment*, CMA Monograph No.179, Oxford & IBH Publishing Company, New Delhi, pp. 205-231.
13. Singh, E.B.K. (1992): *op. cit.,* p. 142.
14. Barker, R., Robert W. Herdt and Beth Rose (1985): "The Rice Economy of Asia", *Resource for the Future*, Washington D.C., p. 53.
15. Singh, E.B.K. (1992): *op. cit.,* p. 142.
16. Bansil, P.C. (1984): *Agricultural Statistics in India*, 3rd edition, Oxford and IBH Publishing Co., New Delhi, p. 53.
17. Sarma, J.S. (1990): "Land Utilisation", in ICAR *Handbook of Agriculture*, New Delhi.
18. a. Percentage of arable land exploited $= \frac{\text{Net area sown}}{\text{Arable land}} \times 100$

    b. Percentage of potential land exploited $= \frac{\text{Net area sown}}{\text{Potential land}} \times 100$
19. Bansil, P.C. (1984): *op. cit.,* p. 193.
20. Government of Manipur: *Agricultural Census of Manipur, 1970-71*, Mimeo, State Agricultural Census Office.
21. Government of Manipur: *Agricultural Census report of Manipur* 1976-77, 1980-81, 1985-86 and 1990-91, Department of Agriculture.
22. NSSO (1997): *Operational Land Holdings in India 1991-92, Salient features*, 48th round (January - December '92), Report No. 407.
23. Bansil, P.C. (1984): *op. cit.,* p. 217.
24. NSS (1988): "A note on some aspects of operational holdings", *Sarvekshana*, July '88.
25. NSSO (1998): "Household Assets and Liabilities as on 30-6-

1991", *Debt and Investment Survey*, 48th Round, Report No. 419.

26. Sawant, S.D. (1997): "Performance of India Agriculture with Special Reference to Regional Variation", *India Journal of Agricultural Economics*, vol. 52, no. 3, July-Sept., p. 363.
27. Singh, E.B.K. (1992): *op. cit.*, p. 104.
28. Bhalla, Sheila (1987): "Trends in Employment in Indian Agriculture, Land and Asset Distribution", *Indian Journal of Agricultural Economics*, vol. 42, no. 4, Oct.-Dec., p. 557.

3

Methodology and Data Sources

3.1 Introduction

The coverage and methodology used to prepare the estimate of capital formation in Indian agriculture are described in the publication of Central Statistical Organisation (CSO)—National Accounts Statistics: Sources and Methods. At the regional level the work has been entrusted to the Department of Economics and Statistics of the respective state. However, only few agriculturally advanced states, under the guidance of CSO, have attempted to estimate the capital formation in agricultural sector of the state by adopting the CSO format. As stated earlier, for the state of Manipur no such attempt has been made till now. Hence the present study is an attempt to estimate a time series on capital formation based on the concepts, definitions, coverage, and methodology adopted by CSO and also analyses other related issues in agricultural sector of Manipur.

3.2 Coverage

Two sets of views on the coverage of economic activities in the analysis of agricultural capital formation are available in the literature. First, those who restrict their analysis to agriculture and allied activities including irrigation system

(Shetty, 1990; Mallick, 1993; Wagle, 1994; Mishra, 1996; Mitra, 1996; Dhawan, 1996a, 1996b, 1997; Misra and Hazell, 1996 & 1997; Dhawan and Yadav, 1997; Misra, 1998; Karmakar, 1998; Sawant et al., 2002). Second, those analysis based on the aggregate of the three sub-sector included in the primary sectors viz. (1) agriculture and allied activities including irrigation system, (2) Forestry and logging, and (3) Fishing (Misra and Chand, 1995; Chand, 2002 & 2001; Gulati and Seema Bathla, 2000 and 2001).

As Sawant et al. (2002) has said, "While analysing temporal trends in agricultural capital formation and particularly the inter-relationship between its two components namely public and private capital formation, forestry and fishing sectors should not be clubbed together. This is because within the sector proportions of public and private capital formation of the two sectors are remarkably different and even temporal trends in total and component-wise capital formation have not been uniform for them."[1] In due consideration of the problem associated with the analysis on aggregate series of the three sub-sector of the primary sector, present study shall restrict to only agriculture and allied activities including irrigation system of Manipur.

As defined in the National Accounts Statistics the scope of agricultural sector for present study includes growing of field crops, fruits, nuts, seeds, vegetable, flowers, tea, coffee and rubber plantations; breeding and rearing of animals and poultry; production of milk, wool, eggs, meal, honey and silk worm cocoons etc; ancillary activities of cultivators such as gur making and extension of agricultural services including operating irrigation services and rental of farm machinery. The operation of irrigation system comprises supply of water through various government channels to the agriculturists.[2] The time period for the present study shall be of 26 years, starting with the year in which Manipur attained full statehood i.e. 1972 till 1997.

3.3 Methodology

The estimate of agricultural capital formation can be prepared by any of the following three methods, provided necessary data are available. They are; (1) Estimation of production of capital goods used in agricultural sector, (2) Estimation of capital expenditure on capital goods, and (3) Accumulation of funds for capital formation i.e. source of finance in the form of savings and loans used for capital formation.[3] As the data on import and export of capital goods with respect to the domestic boundaries of the states and the data on inflow of capital goods from outside the state are not available for the individual state, it is not possible to compile the time series on capital formation with the help of first and third approaches. At the state level, therefore, only the second approach viz. expenditure approach is generally adopted, hence for the present study as well.

The expenditure approach permits the measurement of capital formation within the geographical boundaries of the state, irrespective of the fact whether the agencies involved in capital formation belong to the state or not. This approach has been adopted in the estimation of agricultural capital formation for Tamil Nadu, Karnataka, Andhra Pradesh, Madhya Pradesh, etc.

Capital formation in agriculture sector includes expenditure on land reclamation, land clearance, bunding and other improvements, irrigation and flood control projects including wells and dams, plantations, orchard, and vineyard development, farm buildings, agricultural machinery and equipments such as harvesters, threshers, ploughs, harrows, tractors, etc. and all transport equipments. In other words, gross capital formation in agriculture sector includes only the investment in agriculture. It, however, does not include capital expenditure on construction of warehouses, storage, rural roads and electrification etc. which are termed as investment for agriculture.[4] The exclusion of these activities is a definitional issue but recently more and more researchers have expressed reservation on it. We shall examine it in more detail in our next chapter.

3.4 Sources of Data

The estimates of gross capital formation in agriculture sector are prepared separately for public and private sector as defined in chapter I. The public sector consists of (i) Administrative departments, (ii) Departmental enterprises, (iii) Non-departmental commercial undertakings, and (iv) Local bodies. The estimates of public gross capital formation of the state government viz. administrative departments and departmental enterprises have been compiled from the details available in the budget documents of the state government. The estimates for non-departmental commercial undertakings and local bodies are to be obtained from their respective annual account statement. However, due to the lack of adequate data the present study has excluded this part of public gross capital formation.

From the budget documents capital expenditure incurred by the state government on different budgetary heads have been culled out to arrive at public gross capital formation. There was a change in the classification of capital expenditure under various heads. Since 1974-75 various expenditure heads were grouped into three sectors, viz. general services, social and community services and economic services. Again from 1987-88 onwards the system of accounting was shifted from three digits to four digits classification. In order to maintain uniformity, various budgetary heads of capital expenditure have been adjusted accordingly. The budgetary heads included in capital outlay on agricultural sector as per CSO concept and their correspondence between sub-periods are shown in Table 3.1.

TABLE 3.1

Various Budgetary Heads of Capital Expenditure in Agricultural Sector and Their Equivalence for Pre-1987-88 and Post-1987-88

Sl.No.	*1987-88 onwards*	*Pre 1987-88*
4401	Crop Husbandary	Head 505--agricultural research, storage and marketing
4402	Soil and Water Conservation	Head 506--minor irrigation
4403	Animal Husbandary	Head 510
4404	Dairy Development	Head 511
4701/4711	Major and medium irrigation and flood control projects	Head 532+ Head 533
4702	Minor irrigation	From 506

Note: These expenditure heads includes all the projects developed under North-Eastern Council and Center on the above-cited heads.

Source: Government of Manipur: Annual Budget Documents.

The yearly data on capital expenditure, termed as public gross capital formation in agriculture sector of Manipur, have been compiled at current prices. The series has also been prepared at constant (1980-81) prices, unlike the CSO series due to the lack of other subordinate indices, by deflating the current price series with NSDP deflator.

Capital formation in agriculture sector from private account comprises investments by the corporate sector and household sector. Corporate sector in India is normally categorised into organised and unorganised segments. The organised segment contains big firms primarily in the plantation sector and their estimates of capital formation are available in their accounting books. However, their number is negligible in the state. The present study, therefore, completely excludes this part. Similarly, due to lack of systematic information on data on unorganised components which include all those private cooperatives like sugar, milk, poultry, etc. and other very small cottage agricultural enterprises like dairy, agricultural implements, etc. have also been excluded from the present study. Therefore, the present study shall mainly concentrate on household sector which contribute the largest chunk of private capital formation in agriculture sector of Manipur.

The data on household sector have generally been based on the nationwide decennial survey on household investment behaviour. All India Debt and Investment Survey (AIDIS) 1971-72, Statistical Tables Relating to Capital Expenditure and Capital Formation of Rural Households during the year ended 30th June 1972 brought out by Reserve Bank of India (RBI) and Report on Household Capital Expenditure during 1-7-1991 to 30-6-1992, Debt and Investment Survey, Report No. 437 of NSSO, Government of India are the main sources. Though AIDIS also conducted the similar survey in 1981-82 the report published includes only 17 major states of India. Therefore, the present study shall mainly depend on 1971-72 reports of AIDIS and 1991-92 NSSO report only.

The concepts and definition followed by the two surveys are similar, hence the two reports are comparable. The only difference is that the earlier report is for only rural households whereas the later one includes both rural and urban households. Hence in order to maintain uniformity we shall concentrate only on rural households.

The two surveys give rich information on fixed capital formation by rural households. The data on fixed capital formation by rural households are furnished under three heads, namely - (a) Residential plots and buildings, (b) Farm business, and (c) non-farm business. The estimates of private fixed capital formation in agriculture were arrived at by multiplying per household fixed capital formation in farm business by the number of rural households.

The gross capital formation in private sector is the sum of gross fixed capital formation and change in stocks. However, due to the paucity of data on prices of livestocks, the estimates of change in stocks could not be available. Therefore, the estimates of gross fixed capital formation have been treated as the ad-hoc estimates of gross capital formation in private sector.

In 1971-72, household fixed capital formation in agricultural sector is categorised into eight components:

1. Land Reclamation,
2. Bunding and other land improvements,
3. Orchards and Plantation,
4. Wells,
5. Other irrigation sources,
6. Agricultural implements, machinery and transport equipments, etc.
7. Farm houses and animal sheds, and
8. Other capital expenditure.

In 1991-92 surveys, however, land reclamation and bunding and other land improvements are combined together under one head. Expenditure on transport equipment is taken separately as one of the components of private fixed capital formation. In order to make a comparative analysis of the share of investments of each component in total for both reference points, the present study shall classify only six categories of expenditure heads. The expenditure on land reclamation and bunding and other land improvements are clubbed together under one head. Similarly expenditure on transport equipment in 1991-92 is combined with agricultural implements and machinery head and expenditure on wells and other irrigation sources are also combined together. The six categories of expenditure heads are:

1. Land improvements,
2. Machinery, equipments, and transports,
3. Wells and other irrigation sources,
4. Farm houses and animal sheds,
5. Orchards and plantations, and
6. Other expenditure.

The data available in 1971-72 and 1991-92 surveys were used as benchmark estimates for other years. Since the expenditure incurred on some of the heads in 1971-72 surveys could not be found in 1991-92 surveys, the component wise trend could not be established. The present estimate shall mainly resort to the total fixed expenditure incurred by the rural households in farm business. At the same time due to the lack of adequate subordinate indices/ price indicators to carry forward and backward in order to obtain the time series of private fixed capital formation in agriculture sector of Manipur, the present estimates also take the help of compound growth rate of the two benchmark year to interpolate and extrapolate for the intervening years and to extend beyond 1991-92. Thus, we obtain the series on private fixed capital formation at current prices. To construct the series on private fixed capital formation at 1980-81 prices, NSDP deflator deflates the series on current prices.

3.5 Limitation of Data Base

The database in Manipur is grossly inadequate while the available data are in a very crude form. To that extent the methodology described above suffers from a number of limitations which are listed briefly in the following sub-paras:

1. The budget documents show lump sum amount under various heads of Capital Accounts for which details of expenditure are made on certain assumptions appropriate to the particular Head of Account. This has restricted the construction of capital formation series by types of asset creation and segregated series of gross fixed capital formation and changes in stock on public account. The estimates thus prepared have to be improved based on more reliable database.
2. The share of public sector gross capital formation originating from non-departmental commercial undertakings and local bodies could not be estimated due to lack of adequate database. To that extent public gross capital formation is likely to be underestimated.
3. Unlike the All India Debt and Investment Survey, 1971-72, RBI and Debt and Investment Survey,

1991-92, NSSO, the AIDIS, 1981-82 shows the results of only 17 major states of India. It does not show anything about the remaining small states including Manipur. Thus, the long intervening period (i.e. 20 years) between the two-benchmark surveys may not be a healthy sign. At the same time due to the paucity of data the benchmark figures could not be carried forward and backward with the suitable quantum and price indicators separately for various items included therein. The agriculture sector as a whole in general and gross fixed capital formation on private account in particular are in a very crude form.

4. Due to the non-availability of prices of livestock and its products the series of changes in stock on private account could not be estimated. To that extent estimate of gross capital formation in agriculture and private sector gross capital formation is also deemed to be underestimated.
5. The estimate of gross capital formation at constant prices may not be as accurate as it should be unlike other states and national level, due to the lack of appropriate deflators for different types of asset groups.

In view of the above stated limitations, in the estimates of private capital formation based on secondary data mainly due to the paucity of data, the present study has also attempted to substantiate our findings with the aid of primary data collected through field survey. To evaluate the quantum of private capital formation and to examine its implications on agricultural production a simple random sample survey has been carried out in the four valley districts of Manipur in the year 2000. The survey covered

294 cultivator households. The number of households surveyed in the four districts is given in Table 3.2.

TABLE 3.2

Number of Households Surveyed in the Four Districts

Districts	*No. of households*
Imphal East	97
Imphal West	67
Bishnupur	70
Thoubal	60
Total	294

Source: Field Survey

We were, however, unable to deduce private capital formation resulting from investment on farmers account, as the amount and number of farmers investing on fixed farm assets during the survey period were very low. This has become a constraint in generalizing our findings arising out of survey data. Instead we could only arrive at the stock of capital accumulated at the disposal of the farmers used in agricultural production. The detailed analysis on the quantum and composition of capital stock and its impact on agricultural production shall be discussed in detail in due course of our analysis. It may, however, be pointed out that since the number of cultivator households surveyed in the four districts were not the same, therefore, findings of different districts may not be comparable. But an attempt has been made to visualised the prevalent scene in the four districts.

NOTES

1. Sawant, S.D., Vaidehi Deptardar and Sandhy Mhatre (2002): "Capital Formation and Growth in Agriculture: Neglected

Aspects and Dimensions," *Economic and Political Weekly*, vol. XXXVII, no. 11, March, p. 1068.

2. GOI (1989): *National Accounts Statistics: Sources and Methods*, Central Statistical Organisation, p. 33.
3. Kumar, Baldev (1969): "Capital Formation in Indian Agriculture", *Indian Journal of Agricultural Economics*, vol. 24, no. 4, Oct.-Dec., p. 13.
4. *ibid*., p. 13.

4

Trend of Agricultural Capital Formation in Manipur

4.1 Introduction

The issue of capital formation/investment in agricultural sector has drawn the attention of several scholars and researches in almost every country of the world, more particularly in underdeveloped countries including India, since the work of Nurkse in early 50s. The changing context of world economy and the need for transforming traditional agriculture[1] have further endorsed their endeavour.

In India, the pioneering study on this issue was first conducted by Tara Shukla (1965). The study prepared the estimates of the value of durable physical assets, capital stock, and capital formation for selective years during 1920-21 to 1960-61 at 1950-51 prices both for national level as well as for few selected states. Thereafter the theme of the subject has been discussed in several conferences and seminars and has also received the attention of several individual scholars and researchers. The Indian Society of Agricultural Economics took up this theme for discussion thrice within a short interval of five years, i.e. 1964, 1965, and 1969.

The backdrop of this high interest seems to lie in the country's inability to feed its growing population, continued dependence on PL 480 food imports since the mid-fifties, the food crisis of 1965-67 and the country's development strategy and policy in favour of industrialization at the cost of agriculture.[2] Subsequently the work of estimating capital formation in agriculture has been undertaken by Central Statistical Organization (CSO), Government of India, as a part of National Accounts Statistics (NAS). The CSO estimates of capital formation in agriculture by types of industry and by types of assets at national level are available from 1950-51 onwards at 1980-81 prices and current price as well. But the segregated series by types of institutions viz. Public and Private Sectors are available only from 1960-61 onwards.

Similarly, an attempt has been made to compile the time series of capital formation in agriculture sector of Manipur during 1972-73 to 1997-98 based on the methodology described in the previous chapter. In this chapter, therefore, we shall analyse the direction of movement of estimated series and their growth rate during the said period. But first we shall go through the survey of analysis on CSO series.

4.2 Review of Literature

The CSO series has been widely used by both academic and policy making circles to study the capital formation/investments in Indian agriculture. The whole series has been reproduced along with GDPAGR, AGGGCF and AGGGDP in Table 4.1.

TABLE 4.1

Gross Capital Formation in Indian Agriculture Sector (1950-96)

(Rs. crore at 1980-81 Prices)

Year	*GCF-AGR*	*GDP-AGR*	*col. 2 as percent of col. 3*	*AGG-GCF*	*AGG-GDP*	*col. 5 percent of col. 6*	*col. 2 percent of col. 5*
1	*2*	*3*	*4*	*5*	*6*	*7*	*8*
1950-51	1262	20860	6.050	5746	42871	13.403	21.963
1951-52	1439	21202	6.787	6339	43872	14.449	22.701
1952-53	1332	22084	6.032	4675	45117	10.362	28.492
1953-54	1342	24070	5.575	5143	47863	10.745	26.094
1954-55	1335	24743	5.395	6370	49895	12.767	20.958
1955-56	1582	24377	6.490	7948	51173	15.532	19.904
1956-57	1534	25836	5.937	9517	54086	17.596	16.119
1957-58	1585	24523	6.463	9964	53432	18.648	15.907
1958-59	1510	27281	5.535	8390	57487	14.595	17.998
1959-60	1294	26883	4.813	9503	58745	16.177	13.617
1960-61	1668	28841	5.783	11775	62904	18.719	14.166
1961-62	1670	28748	5.809	10813	64856	16.672	15.444
1962-63	1848	28131	6.569	12692	66228	19.164	14.560
1963-64	2000	28658	6.979	13535	69581	19.452	14.777
1964-65	2128	31619	6.730	14249	74858	19.035	14.934
1965-66	2308	27360	8.436	14743	72122	20.442	15.655
1966-67	2346	26734	8.775	15455	72856	21.213	15.180
1967-68	2589	31298	8.272	14891	78785	18.901	17.386
1968-69	2694	31190	8.637	13249	80841	16.389	20.334
1969-70	2871	33451	8.583	15102	86109	17.538	19.011
1970-71	2758	35930	7.676	16550	90426	18.302	16.665
1971-72	2924	34973	8.361	17941	91339	19.642	16.298
1972-73	3180	33005	9.635	17636	91048	19.370	18.031
1973-74	3208	35786	8.964	20007	95192	21.018	16.034
1974-75	2975	34800	8.549	20729	96297	21.526	14.352

contd...

1	*2*	*3*	*4*	*5*	*6*	*7*	*8*
1975-76	3388	39740	8.525	22908	104968	21.824	14.790
1976-77	4258	37323	11.409	22498	106280	21.169	18.926
1977-78	4073	41994	9.699	22415	114219	19.625	18.171
1978-79	5246	42831	12.248	28144	120504	23.355	18.640
1979-80	5215	37108	14.054	27334	114236	23.928	19.079
1980-81	4636	42466	10.917	25794	122427	21.069	17.973
1981-82	4503	45145	9.975	32437	129889	24.973	13.882
1982-83	4590	44570	10.298	32496	133915	24.266	14.125
1983-84	4101	49753	8.243	31920	144865	22.034	12.848
1984-85	4549	49702	9.153	33187	150433	22.061	13.707
1985-86	4325	49855	8.675	38654	156566	24.689	11.189
1986-87	4011	48995	8.187	38405	163271	23.522	10.444
1987-88	4414	49258	8.961	37679	170322	22.122	11.715
1988-89	4346	57940	7.501	44761	188461	23.751	9.709
1989-90	4353	58568	7.432	44263	201453	21.972	9.834
1990-91	4594	60991	7.532	49886	212253	23.503	9.209
1991-92	4729	59398	7.962	46718	213983	21.833	10.122
1992-93	5372	63327	8.483	52131	225240	23.503	10.305
1993-94	5031	65762	7.650	51330	239145	21.464	9.801
1994-95	6256	69206	9.040	71422	257700	27.715	8.759
1995-96	6961	66863	10.411	76820	276132	27.820	9.061
1996-97	6999	72362	9.672	74216	296845	25.002	9.431

Source: National Accounts Statistics of India: 1950-51 to 1996-97, 3rd Edition, EPW Research Foundation, Mumbai (India) Oct., 1998 and National Accounts Statistics, CSO various issues.

Note:
1. GCFAGR = Gross Capital Formation in Agriculture sector.
2. GDPAGR = Gross Domestic Product originating from agriculture.
3. AGGGCF = Aggregate Gross Capital Formation for the economy as a whole.
4. AGGGDP = Aggregate Gross Domestic Product for the economy as a whole.

The series shows that gross capital formation in Indian agriculture (GCFAGR) at 1980-81 prices has been steadily rising over the period 1950-78 from a meager level of Rs. 1262 crore in 1950-51 to Rs. 5246 crore in 1978-79. It then declined persistently up to 1986-87 touching Rs. 4353 crore and then remained almost stagnant up to 1989-90. However, from 1990-91 it has been gradually rising touching the figure of Rs.6999 crore by 1996-97. The annual compound growth rate of GCFAGR, as shown in Table 4.2, during the entire period 1950-96 is 3.703 percent but growing at decelerating rate. At sub-period level it grew at an annual compound growth rate of 1.221 percent in the fifties, 6.558 percent in the sixties, 7.571 percent in the seventies but no discernible growth rate could be observed in the eighties. Thus, there was a noticeable stagnation in GCFAGR in the eighties after an accelerating high growth rate during the previous decade of seventies. In the nineties it, however, picked up again with an annual growth rate of 8.130 percent.

In relation to the real GDP originating from agricultural sector, GCFAGR at 1980-81 prices formed about 6 percent upto the early part of sixties, about 8 percent during the period 1965-75, then increased to a peak of 14 percent in 1979-80. It then declined gradually till 1993-94 touching the figure of 7.65 percent only to pick up again slightly in the last three years to about 9 percent. Unlike the agriculture sector, the rate of gross capital formation in the economy as a whole (AGGGCF) has been steadily rising since 1950s though there is some year-to-year fluctuation. The share of AGGGCF in aggregate gross domestic product of the economy (AGGGDP) increased from 13.4 percent in 1950-51 to the height of 27.82 percent in 1995-96.

On the other hand, the share of GCFAGR in AGGGCF declined in the fifties from 28.49 percent in 1953-54 to 13.62 percent in 1959-60, then increased to 20.33 percent in 1968-69. Since then with exception of 1976-80, when it remained relatively stagnant to around 18.5 percent, it continued to decline touching the lowest level of 8.76 percent in 1994-95. Thus, unlike the aggregate gross capital formation in the economy, GCFAGR declined in the 80s. This has been further endorsed when we look at the annual compound growth rate of AGGGCF during the corresponding period.

TABLE 4.2

Annual Compound Growth Rate

Period	*GCF-AGR*	*AGG-GCF*	*GFC-FAGR*	*AGG-GFCF*	*GDP-AGR*	*AGG-GDP*	*PUB-GCF*	*PVT-GCF*
1950-96	3.703	5.499	3.655	5.481	2.573	4.063	2.192*	3.799*
1950-59	1.221	7.767	0.876	7.447	2.932	3.681	-	-
1960-69	6.558	3.01	6.485	3.860	1.272	3.287	2.540	8.484
1970-79	7.571	5.81	6.058	5.492	1.941	3.448	9.597	6.628
1980-89	-0.681	5.326	-0.750	5.600	3.137	5.407	-4.373	1.285
1990-96	8.130	9.344	8.122	8.936	3.040	6.073	2.270	9.672

Source: Same as Table 4.1.
Note: * For the period 1960-96.

The annual compound growth rate of AGGGCF during 1950-96 is 5.499 percent, which grew at 7.767 percent in the fifties, 3.01 percent in the sixties, 5.81 percent in the seventies, 5.326 percent in the eighties and a phenomenally high growth rate of 9.344 percent in the nineties. While the annual compound growth rate of GDPAGR during 1950-96 is 2.573 percent only. It grew at 2.932 percent in the fifties, 1.272 percent in the sixties, 1.941 percent in the seventies, 3.137 percent in the eighties and 3.04 percent in the nineties.

Table 4.3 shows the trend of gross fixed capital formation in agriculture sector (GFCFAGR), changes in inventory stocks (CIS) i.e. the difference between gross and fixed capital formation in agriculture sector, gross fixed capital formation for the economy as a whole (AGGGFCF), and the percentage of GFCFAGR to AGGGFCF. The GFCFAGR has increased from Rs. 1224 crore in 1950-51 to Rs. 4537 crore in 1980-81, then declined to Rs. 3798 crore in 1986-87 and remained almost stagnant at that level till 1989-90. From 1990-91 onwards it has been persistently rising touching Rs. 6856 crore in 1996-97. However, when we look at the aggregate series for the economy as a whole AGGGFCF has been continuously rising with some minor variation from a mere Rs. 4891 crore in 1950-51 to Rs. 69,142 crore in 1996-97. The percentage share of GFCFAGR in AGGGFCF has declined from 25 percent in 1950-51 to almost 10 percent in 1996-97 except during seventies when its share was relatively stable at about 19 to 20 percent.

TABLE 4.3

Gross Fixed Capital Formation in Indian Agriculture Sector (1950-96)

(Rs. crore at 1980-81 prices)

Year	*GFCFAGR*	*CIS*	*AGGGFCF*	*col. 2 as percent of col. 4*
1	*2*	*3*	*4*	*5*
1950-51	1224	38	4891	25.026
1951-52	1436	3	5332	26.932
1952-53	1244	88	4883	25.476
1953-54	1414	-72	5361	26.376
1954-55	1367	-32	5980	22.860
1955-56	1500	82	7111	21.094
1956-57	1489	45	8109	18.362
1957-58	1473	112	8584	17.160
1958-59	1413	97	8029	17.599
1959-60	1288	6	8292	15.533
1960-61	1585	83	9793	16.185
1961-62	1665	5	9572	17.394
1962-63	1804	44	11037	16.345
1963-64	1939	61	11963	16.208
1964-65	2103	25	12660	16.611
1965-66	2258	50	13599	16.604

contd...

1	*2*	*3*	*4*	*5*
1966-67	2313	33	13239	17.471
1967-68	2580	9	13272	19.439
1968-69	2558	136	12756	20.053
1969-70	2754	117	13578	20.283
1970-71	2625	133	13762	19.074
1971-72	2767	157	14351	19.281
1972-73	2938	242	16277	18.050
1973-74	2902	306	15767	18.406
1974-75	2709	266	15515	17.461
1975-76	2935	453	18117	16.200
1976-77	3646	612	18859	19.333
1977-78	3744	329	19927	18.789
1978-79	4246	1000	22150	19.169
1979-80	4440	775	21653	20.505
1980-81	4537	99	23617	19.211
1981-82	4346	157	26408	16.457
1982-83	4409	181	28607	15.412
1983-84	3957	114	28708	13.784
1984-85	4287	262	30560	14.028
1985-86	4068	257	31781	12.800
1986-87	3798	213	34231	11.095
1987-88	4219	195	35848	11.769
1988-89	4260	86	37954	11.224
1989-90	4191	162	40062	10.461
1990-91	4459	135	44092	10.113
1991-92	4667	62	45131	10.341
1992-93	5260	112	47698	11.028
1993-94	5005	26	52318	9.566
1994-95	6110	146	62793	9.730
1995-96	6777	184	66388	10.208
1996-97	6856	143	69142	9.916

Sources: Same as Table 4.1.

Note: GFCFAGR = Gross Fixed Capital Formation in Agriculture.
CIS = Changes in Inventory Stock in Agriculture.
AGGGFCF =Aggregate Gross Fixed Capital Formation in the Economy

An examination on the relation between gross capital formation and gross fixed capital formation in agriculture revealed that the share of changes in inventory stock within the sector during the seventies have generally been higher than those of two previous decades. It was highest in 1978-79, when it reached Rs.1000 crore, which is in coincidence with the highest gross capital formation in agriculture. It, however, declined tremendously during the two subsequent decades of eighties and nineties.

The annual compound growth rate of GFCFAGR during the period 1950-96 was 3.655 percent per annum. It grew at 0.876 percent in the fifties, 6.485 percent in the sixties and 6.058 percent in the seventies which is also accelerating. However, during the eighties there is no discernible growth similar to that of GCFAGR. But in the nineties it again grew at the rate of 8.122 percent per annum. For the economy as a whole (AGGGFCF), it is growing at the rate of 5.481 percent per annum during the period 1950-96. It grew at 7.447 percent in the fifties, 3.86 percent in the sixties, 5.492 percent in the seventies, 5.6 percent in the eighties and 8.936 percent in the nineties. Thus, unlike the rest of the economy there is a noticeable decline in the rate of asset formation in agriculture during the eighties.

The GCFAGR series segregated by types of institutions viz. Public and Private, is also available but only from 1960-61 onwards. The series, given in Table 4.4, shows that the public gross capital formation (PUBGCF) in agriculture increased from mere Rs. 582 crore in 1960-61 to Rs. 1796 crore in 1980-81 and remained stagnant around that level till 1984-85 which declined thereafter till 1991-92 touching the lowest level of Rs. 1002 crore. It then picked up somewhat till 1994-95 but declined again to Rs. 1132 crore in 1996-97. On the other hand private gross

TABLE 4.4

Gross Public and Private Capital Formation in Indian Agriculture Sector at 1980-81 Price

(Rs. crore)

Year	*GCFAGR*	*PUBGCF*	*PVTGCF*	*col. 3 as per-cent of col. 2*
1	*2*	*3*	*4*	*5*
1950-51	1262			
1951-52	1439			
1952-53	1332			
1953-54	1342			
1954-55	1335			
1955-56	1582			
1956-57	1534			
1957-58	1585			
1958-59	1510			
1959-60	1294			
1960-61	1668	589	1079	35.312
1961-62	1670	600	1070	35.928
1962-63	1848	694	1154	37.554
1963-64	2000	725	1275	36.250
1964-65	2128	765	1363	35.949
1965-66	2308	798	1510	34.575
1966-67	2346	696	1650	29.668
1967-68	2589	688	1901	26.574
1968-69	2694	775	1919	28.768
1969-70	2871	775	2096	26.994
1970-71	2758	789	1969	28.608
1971-72	2924	851	2073	29.104
1972-73	3180	1049	2131	32.987
1973-74	3208	993	2215	30.964

1	*2*	*3*	*4*	*5*
1974-75	2975	919	2056	30.891
1975-76	3388	1041	2347	30.726
1976-77	4258	1378	2880	32.363
1977-78	4073	1534	2539	37.663
1978-79	5246	1697	3549	32.348
1979-80	5215	1772	3443	33.979
1980-81	4636	1796	2840	38.740
1981-82	4503	1781	2722	39.551
1982-83	4590	1742	2848	37.952
1983-84	4101	1711	2390	41.722
1984-85	4549	1674	2875	36.799
1985-86	4325	1520	2805	35.145
1986-87	4011	1425	2586	35.527
1987-88	4414	1458	2956	33.031
1988-89	4346	1362	2984	31.339
1989-90	4353	1156	3197	26.556
1990-91	4594	1154	3440	25.120
1991-92	4729	1002	3727	21.188
1992-93	5372	1061	4311	19.751
1993-94	5031	1153	3878	22.918
1994-95	6256	1316	4940	21.036
1995-96	6961	1268	5693	18.216
1996-97	6999	1132	5867	16.174

Source: Same as Table 4.1.

Note: 1. GCFAGR = Gross Capital Formation in Agriculture Sector.
2. PUBGCF = Public Gross Capital Formation.
3. PVTGCF = Private Gross Capital Formation.

capital formation (PVTGCF) increased from Rs. 1079 crore in 1960-61 to Rs. 5867 crore in 1996-97 with the exception of late seventies to mid-eighties when there was a slight fall in the rate of private asset formation. The comparative movement of public and private capital formation can also be seen from the graph given as Figure 4.1.

Figure 4.1

Trends of Public, Private and Gross Capital Formation in Indian Agriculture during 1960 to 1996
(at 1980-81 prices)

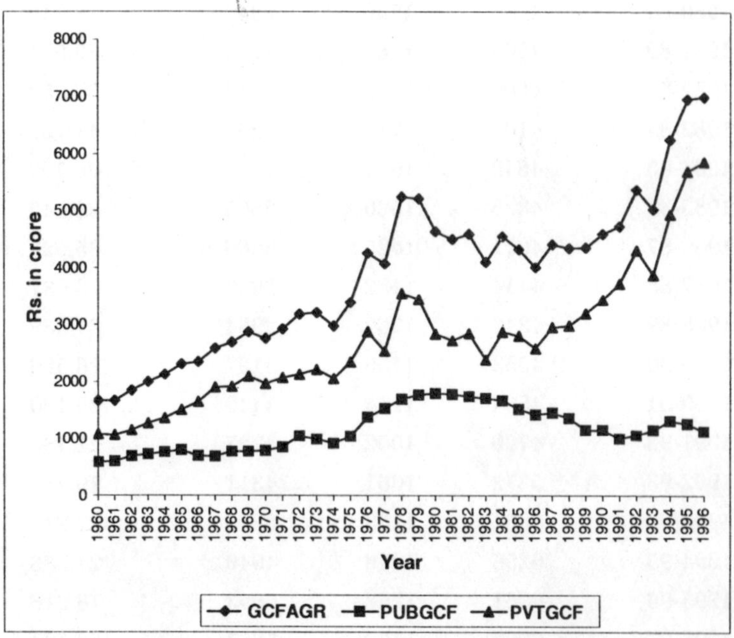

The graph shows that from mid eighties onwards PUBGCF has declined tremendously while the opposite is true in the case of PVTGCF. That is to say that the increase in GCFAGR since mid-eighties is mainly due to rise in PVTGCF. This is further endorsed when we look at the annual compound growth rate of the PUBGCF and PVTGCF during different sub-periods.

The annual compound growth rate of PUBGCF, given in Table 4.2, during the period 1960-96 is growing at a decelerated rate of 2.192 percent. It grew at 2.54 percent in the sixties, 9.597 percent in the seventies, -4.373 percent in the eighties and 2.27 percent in the nineties. The negative growth rate in the eighties though at decelerated rate indicates that PUBGCF is declining. On the contrary, PVTGCF during the entire period is growing at the rate of 3.799 percent per annum. Except during the decade of seventies, the annual compound growth rate of PUBGCF is higher than that of PVTGCF. Over and above PVTGCF is growing at an accelerated rate, though at 10% level of significance, during the decade of seventies and eighties. This indicates that the decline in GCFAGR during eighties is mainly due to the decline in PUBGCF.

There has been an intense debate in the country, both in academic and policy making circles, that gross capital formation in Indian agriculture - public capital formation in particular has been either stagnating or falling since the beginning of eighties and it does not augur well for the nation's food security and overall growth of agriculture sector (Shetty, 1990; Gandhi, 1990 & 96; Kumar, 1992; Mallick, 1993; Mishra and Chand, 1995; Mitra, 1996; Dhawan, 1996a, 1996b, & 1998, Banerjee, 1996; Mani et al 1996; Alagh, 1997; Dhawan and Yadav, 1997 etc.). It has also been acknowledged officially (Economic Survey, 1993-94). They have identified various factors/ reasons responsible for this decline.

According to Shetty (1990), the decline in agricultural capital formation, both in public and private sector, during eighties as compared to the previous decades was mainly due to decline in public sector investment as revealed by decline in expenditure incurred for agriculture and irrigation as a proportion of aggregate expenditure of the center and states and resultant loss of momentum in private sector investment. Reduced public sector investment in agriculture combined with an unattractive growth horizon, adverse terms of trade, low per capita income growth and inadequate growth in savings may have depressed household's ability and incentive to invest in agriculture.[3] Similarly, Mallick (1993) has also concluded that the decline in agricultural capital formation in the eighties is mainly due to decline in real public investment which may be affecting private capital formation in agriculture though hypothetically several factors like low domestic saving, low productivity, adverse terms of trade, the role of institutional credit and structural factors like farm size are also operating to some extent.[4]

The decline in real public investment, on the other hand, is mainly due to increasing flow of resources to current expenditure on subsidies for fertilizers, irrigations, electricity, credit and other agricultural inputs, rather than investment on fixed assets. Institutional credit induced the farmer to invest more, particularly in the context of technological changes. However, since the mid sixties i.e. after the green revolution, no further technological break through has taken place that has acted as constraint in the expansion of the role of institutional credit. On the other hand, the average size of farms has been declining over the year due to demographic pressure which limits the growth of private investment as they have not been profitable.[5]

The decline in agricultural investment, more particularly private sector investment during 1980-86, are mainly attributed to squeezing of rural savings as well as lowering of net commercial bank and cooperative credit to agriculture in real term.[6] While Mitra (1996) attributed the decline in public investment in Indian agriculture to fall in government's expenditure (outlay) on agriculture and allied activities, irrigation and rural development and other special area programmes.[7] In brief, with the exception of the study by Gandhi (1990 and 1996) various factors that have been put forth behind the observed trend of falling agricultural gross capital formation in general and public capital formation in particular that started since late seventies revolve around the issue of increase in subsidies on various inputs to agriculture sector, adverse terms of trade, reduction in resources flow from center to states and decline in capital expenditure (outlay) by the state.

Falling real public investment in agriculture is the major source of decline in capital formation in Indian agriculture sector. Agricultural investment policy followed during the decade has come under severe criticism. Kumar (1992) termed the fall in public investment in agriculture as 'neglect of agriculture'. He attributed this bias against agriculture to the government policy.[8] Mishra and Chand (1995) and Mishra (1996) attributed it to political economy of agricultural policy. The politically powerful interest seeking farmers' groups that had emerged towards the end of the seventies have increasingly come to determine the pattern of public expenditure on agriculture. Their main priority was to increase the share of current account expenditure to meet the demand for production subsidies and also to finance private sector capital formation by institutional term loans and capital subsidies, leaving only the residual for public capital formation.[9] Public financing of private investments rose to 60 percent in post green revolution period leaving not much room for the state for public investments. On the other hand pressure put forth by environmental groups against the construction of major and medium irrigation systems and inter-state water disputes had adverse effect on public sector investment, which has been largely for development of irrigation facilities.[10]

As stated earlier, public investment is mainly meant for infrastructural development. It, therefore, plays a crucial role in the long-term growth of output. Accordingly, decline in its level would have an adverse impact on the growth of agricultural output. While acknowledging the importance of public investment for the growth of output some researchers have expressed the reservation on the CSO series, which relied more on indirect than direct methods (Mishra and Chand, 1995; and Mitra, 1997).

The CSO series on public investment is deemed to be underestimated as it counts only the investment in agriculture, which is mainly dominated by investment in medium and medium irrigation systems. The series does not include the investment in infrastructural activities like rural roads, rural electrification, rural markets, godowns, storages, etc. which are termed as investment for agriculture.[11] These activities also play a crucial role indirectly, if not directly, in growth and development of this sector. Some researchers asserted that a decline in public investment in agriculture need not necessarily imply a decline in public investment for agriculture. This has instigated the researchers in search of a more comprehensive series on public capital formation in agriculture that includes all the important items relevant to agriculture sector. Thus, in recent years some of them have attempted to re-estimate the public investment in agriculture with coverage wider than conventional CSO coverage and attempted to examine various issues related with it.

Chand (2000) is the first to attempt to re-estimate the public investment in agriculture by taking 23 heads of capital expenditure[12] reported in the Finance Accounts of the central and state governments for the period 1974-75 to 1996-97. His broad series also revealed that capital expenditure on agriculture in real term for the country as a whole declined throughout the period under study and little earlier than the decline in the CSO series. He, thus, concluded, "it is not correct to say that public investment for agriculture which includes all major heads like rural roads, rural electrification, storage, warehousing etc. has not declined even though CSO series has declined. Further, the rate of decline in capital outlay on irrigation and allied heads was lower compared to the rate of decline in capital expenditure on other agricultural heads and in CSO series."[13]

Similarly Gulati and Seema Bathla (2000) examined critically from the very concept and estimation procedures followed in Indian system of National Accounts vis-à-vis, the UN system by taking into account the 11 heads of capital expenditure[14] apart from investments in major, medium and minor irrigation works that have been included in the budgetary classification of agriculture and allied sector. Their broad series also revealed that public sector gross capital formation in agriculture sector is either stagnating or declining since mid eighties. However, according to them, though the decline in public capital formation definitely affects the growth of output, it is not as alarming as is sometimes made out to be. They concluded with a speculation, "if India had not neglected the public sector gross capital formation in agriculture (GCFA) during 1980s and early 1990s the growth in gross domestic product originating from agriculture (GDFA) during 1990s would have been above 4 percent per annum. This would have placed India comfortably on an overall growth path of more than 7 percent per annum."[15]

Having discussed the time series, growth rate and also the critical review of literature on the decline in capital formation in general and public capital formation in particular in Indian agriculture in the eighties we shall now extend our analysis at the regional level for the state of Manipur. But first we shall look at the structure of public and private capital formation in agriculture sector of Manipur.

4.3 Structure of Public Capital Formation in Manipur

As described in the Chapter III, public sector capital formation in agriculture is measured as capital expenditure incurred by central and state government and North-East Council on six different heads reported in the budgetary documents. The list of capital expenditure heads included in the compilation of public sector capital formation is given in the Table 3.1 of the previous chapter. It is found that, throughout the period irrigation and flood control accounted the major chunk of the total public sector capital formation. Its share rose from 56 percent in 1972-73 to almost 99% in 1997-98 except during late eighties to early nineties when it declined marginally. In absolute term (at current price) it increased from Rs.72.28 lakh in 1972-73 to Rs. 4108 lakh in 1997-98 with some minor fluctuation. However, at 1980-81 prices, it increased from Rs. 1.39 lakh in 1972-73 to Rs. 14.12 lakh in 1991-92. Since then it remained stagnant at around Rs. 11 lakh till 1997-98.

Contrary to that the share of capital expenditure on non-irrigation related heads, viz. crop husbandry, soil and water conservation, animal husbandry, and dairy development has been declining considerably. In absolute term (at 1980-81 prices) it increased from Rs. 1.08 lakh in

1972-73 to Rs.3.05 lakh in 1978-79 then remained stagnant at around Rs. 1.5 lakh between 1979 to 1986 and at around Rs. 3.5 lakhs between 1987 to 1991. From 1992 onwards it declined to less than Rs. 0.5 lakh till 1997.

4.4 Structure of Private Capital Formation in Manipur

Agricultural capital formation on private account is quite a varied basket of investment goods. The share of each item varied from one benchmark to another i.e. AIDIS 1971-72, RBI to DIS 1991-92, NSSO. The percentage share of each item of private capital formation and the bar chart is shown in Table 4.5.

TABLE 4.5

Percentage Distribution of Fixed Capital Expenditure of Rural Households in Agriculture during 1971-72 and 1991-92 in Manipur

Sl.No.	*Scheme / Source*	*1971-72*	*1991-92*
1.	Land improvements	38.18	1.43
2.	Machinery, Equipments & Transport	7.82	0.29
3.	Well and other irrigation sources	25.38	--
4.	Farm houses, barns & Animal shed	4.49	4.60
5.	Orchards and Plantation	1.1	--
6.	Other Capital Expenditure	23.03	93.68
7.	Total	100	100

Source: 1. RBI (1978): "Statistical Tables Relating to Capital Expenditure and Capital Formation of Rural Households during the year ended 30th June 1972", AIDIS 1971-72.

2. NSSO (1998): "Household Capital Expenditure during 1-7-91 to 30-6-92", Debt and Investment Survey, 48th round.

Accordingly in 1971-72 survey, 'land improvements' which include land reclamation, bunding and other land improvement, is the single most important item accounting about Rs.17 out of Rs.44 spent by every household in asset creation. That is to say that land improvement constitute around 38 percent of the total capital formation on private account in 1971-72. Secondly, irrigation related capital formation covering wells and other irrigation sources accounts for about Rs.11 per rural household which constitute about 25 percent. The third largest is other capital expenditure including furniture, fixture and any cost borne on any other item of farm business like preparing hedges for erecting fences around the field etc. which were undertaken for permanent protection or improvements of the farm accounted for about Rs.10 per rural household. It constitutes about 23 percent.

The shares of each item reported in Debt and Investment survey (DIS) 1991-92 have been completely changed from the earlier one. In 1991-92, other expenditure accounts Rs.326 out of total expenditure on asset creation of Rs. 348 per rural household. That is to say that other expenditure accounted around 94 percent of the total capital formation on private account during 1991-92. It is followed by farm houses, barns and animal sheds with 4.6 percent. The expenditure on wells and other irrigation sources and orchards and plantations leading to capital formation has not been reported at all.

Capital formation in agriculture on private account per rural household in 1971-72 was rather low amounting to about Rs.44 out of total gross capital expenditure of Rs.108 incurred by per rural household. It means that only 41 percent of the gross capital expenditure a rural household was spent on asset creation in 1972-73. However, in 1991-92 out of Rs. 373 gross capital expenditure incurred by per rural household, Rs. 348 was devoted to capital formation. That is to say that about 93 percent of gross capital expenditure per rural household were utilized for capital formation.

Bar Chart

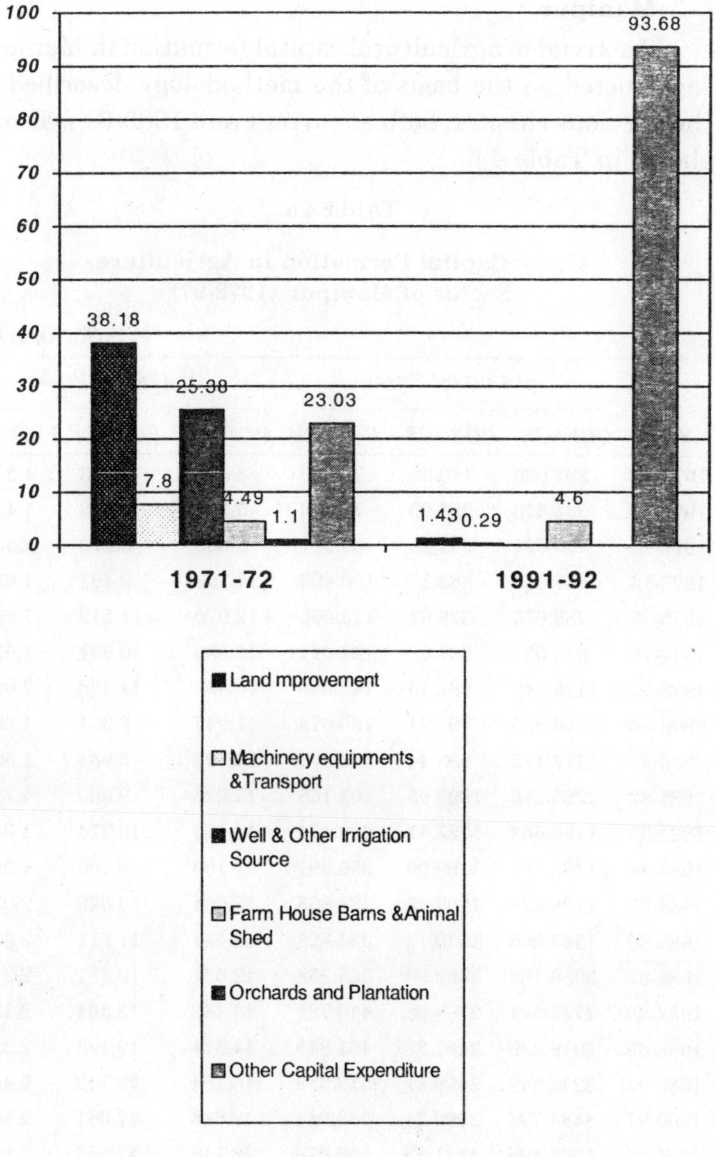
100
90
80
70
60
50
40
30
20
10
0
38.18
7.8
25.38
4.49
1.1
23.03
1.43
0.29
4.6
93.68
1971-72
1991-92
Land mprovement
Machinery equipments &Transport
Well & Other Irrigation Source
Farm House Barns &Animal Shed
Orchards and Plantation
Other Capital Expenditure

4.5 Trend of Agricultural Capital Formation in Manipur

The trend of agricultural capital formation in Manipur constructed on the basis of the methodology described in the previous chapter, both at current and 1980-81 prices is shown in Table 4.6.

TABLE 4.6

Gross Capital Formation in Agriculture Sector of Manipur (1972-97)

(Rs. in lakh)

	At Current Prices			*At 1980-81 Prices*		
Year	*GCFAGR*	*PUBGCF*	*PVTGCF*	*GCFAGR*	*PUBGCF*	*PVTGCF*
1972-73	201.000	130.32	70.680	3.806	2.468	1.338
1973-74	303.378	223.90	79.478	5.390	3.978	1.412
1974-75	396-621	307.25	89.371	6.685	5.179	1.506
1975-76	663.666	563.17	100.496	11.153	9.464	1.689
1976-77	788.675	675.67	113.005	12.970	11.112	1.858
1977-78	815.051	687.98	127.071	12.820	10.821	1.999
1978-79	1125.049	982.16	142.889	16.493	14.398	2.095
1979-80	958.385	797.71	160.675	11.247	9.361	1.886
1980-81	1179.115	998.44	180.675	11.791	9.984	1.807
1981-82	1293.215	1090.05	203.165	11.276	9.505	1.771
1982-83	1456.564	1228.11	228.454	12.185	10.274	1.911
1983-84	1455.981	1199.09	256.891	11.159	9.190	1.969
1984-85	1798.198	1509.33	288.868	12.664	10.629	2.034
1985-86	1953.965	1629.14	324.825	13.447	11.211	2.236
1986-87	2058.188	1692.93	365.258	12.621	10.381	2.240
1987-88	2777.344	2366.62	410.724	14.439	12.304	2.135
1988-89	2949.239	2487.39	461.849	14.374	12.123	2.251
1989-90	3218.549	2699.21	519.339	15.194	12.743	2.451
1990-91	4484.724	3900.74	583.984	19.615	17.061	2.554
1991-92	5098.496	4441.82	656.676	20.148	17.553	2.595

contd...

1992-93	3719.766	2981.35	738.416	13.947	11.178	2.769
1993-94	4289.982	3459.65	830.332	13.717	11.062	2.655
1994-95	4713.318	3779.63	933.688	14.308	11.474	2.834
1995-96	5284.430	4234.52	1049.910	15.235	12.207	3.028
1996-97	5894.738	4714.14	1180.598	16.291	13.028	3.263
1997-98	5488.665	4161.11	1327.555	14.433	10.942	3.491

Source: 1. Government of Manipur: Annual Budget Documents of the state
2. Same as Table 4.5

Note: 1. GCFAGR = Gross Capital Formation in Agricultural Sector of Manipur.
2. PUBGCF = Public Sector Gross Capital Formation in Agricultural Sector of Manipur.
3. PVTGCF = Private Sector Gross Capital Formation in Agricultural Sector of Manipur.

The Table shows that gross capital formation (GCFAGR) at current prices is persistently rising till 1991-92 when it touched Rs.5098.5 lakh from a mere level of Rs. 201 lakh in 1972-73. It, however, suddenly slip down to Rs. 3719.77 lakh in 1992-93 only to recover again gradually to Rs. 5488.67 lakh in 1997-98. On the other hand, GCFAGR at 1980-81 prices shows a wide range of fluctuation over the entire period covered. It has been steadily rising till late seventies from a meager level of Rs. 3.81 lakh in 1972-73 to Rs.16.49 lakh in 1978-79. It then suddenly slip down to Rs. 11.25 lakh in 1979-80 and remained almost stagnant at that level till 1986-87, then gradually rised to peaked at Rs. 20.15 lakh in 1991-92 but slip down again to around Rs. 14 lakh in 1992-93 and remained almost stagnant till 1997-98.

An examination on annual compound growth rate as shown in Table 4.7 during the period 1972-73 to 1997-78 revealed that real GCFAGR is growing at a decelerating rate of 3.37 percent. It grew at an extraordinarily high but decelerating rate of 19.45 percent in the seventies, 3.24 percent in the eighties and -3.396 percent in the sub-period of nineties. The negative growth rate recorded in the nineties indicated that GCFAGR is declining at an accelerated rate (significant at 10% level of significance). Thus, there is a noticeable decline in real GCFAGR in the nineties after the positive growth rate during the two previous decades.

TABLE 4.7

Annual Compound Growth Rate

Period	*PUBGCF*	*GCFAGR*	*NSDP*	*NSDPAGR*
1972-97	3.513	3.371	4.771	1.857
1972-79	24.099	19.952	5.902	6.551
1980-89	3.213	3.239	4.816	1.047
1990-97	-4.989	-3.396	5.219	2.962

Source: Table 4.6 & 4.8.

When we look at the segregated series of public and private sector capital formation, we shall have to concentrate mainly on public sector series since the private sector series has been derived by interpolating and extrapolating with simple compound growth rate of the figure reported in the two benchmark surveys of AIDIS 1971-72 and DIS 1991-92. It can also been seen from the graph of GCFAGR, PUBGCF and PVTGCF at current prices and1980-81 prices given in Figure 4.2 and 4.3 respectively.

The public capital formation at current prices shows a persistently rising trend with some minor fluctuation during the period 1972-73 to 1991-92 which increases from Rs. 130.32 lakh in 1972-73 to Rs. 4441.32 lakh in 1991-92. It, however, declined to Rs. 2981.35 lakh in 1992-93 only to recover again marginally thereafter touching the level of Rs. 4161.11 lakh in 1997-98. On the other hand, PUBGCF at 1980-81 prices increases from a meager level of Rs. 2.47 lakh in 1972-73 to Rs. 14.4 lakh in 1978-79, then declined

Figure 4.2

Trends of Public, Private and Gross Capital Formation in Agriculture Sector of Manipur during 1972 to 1997 (at current prices)

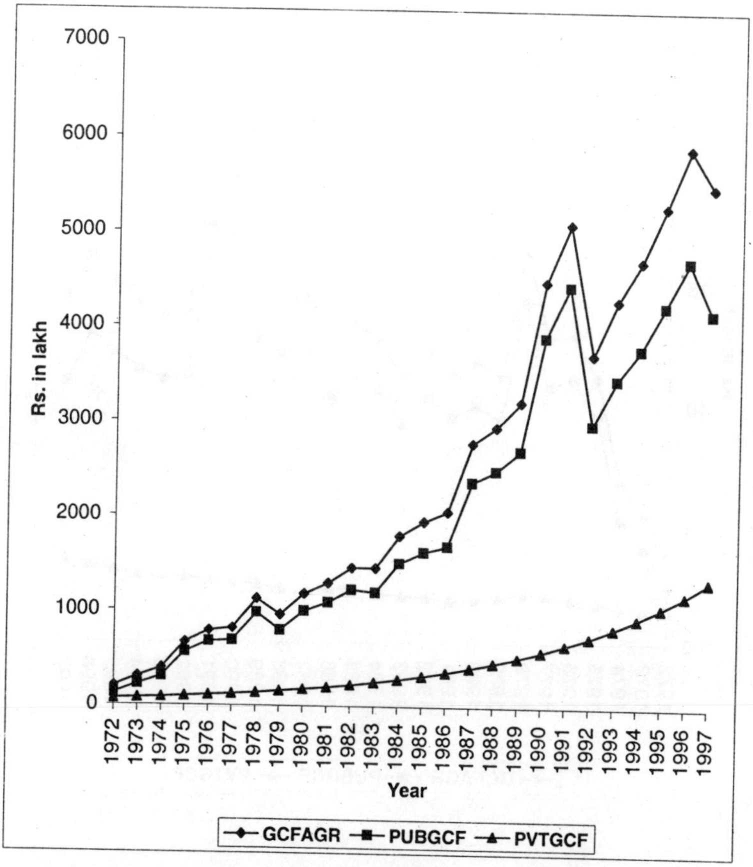

to Rs. 9.36 lakh in 1979-80 and remained almost stagnant till, 1986-87. Thereafter it gradually increased to Rs. 17.55 lakh in 1991-92 only to slump down again to Rs. 11.98 lakh in 1992-93 and remained stagnant at that level, with some fluctuation, till 1997-98.

Figure 4.3

Trends of Public, Private and Gross Capital Formation in Agriculture Sector of Manipur during 1972 to 1997 (at 1980-81 prices)

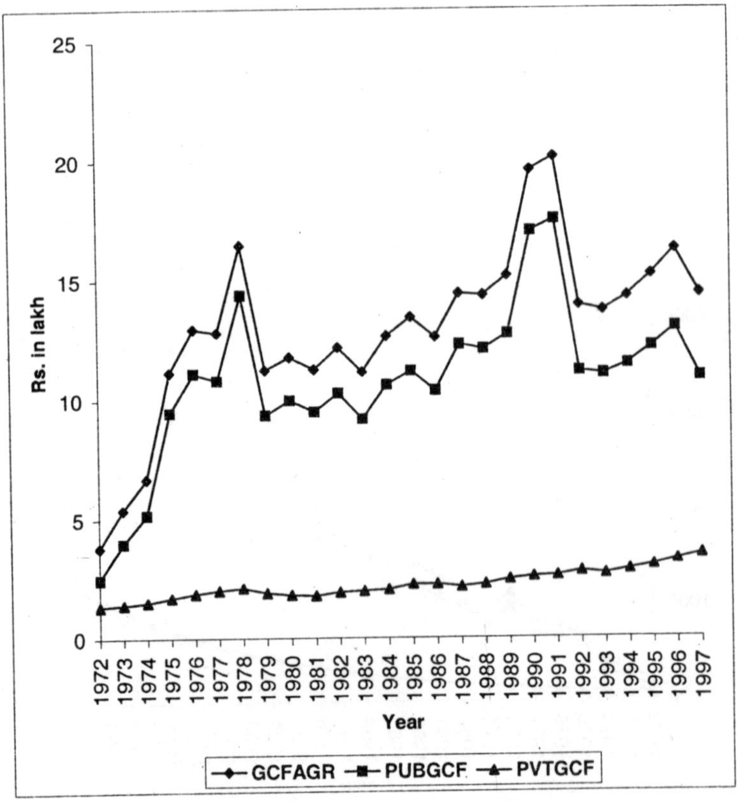

The picture would be more clear when we look at the annual compound growth rate. During the period 1972-73 to 1997-98 PUBGCF is growing at a decelerated rate of 3.51 percent per annum. It grew at an extraordinarily high but decelerated rate of 24.1 percent in the seventies, 3.21 percent in the eighties and -4.989 percent in the sub-period of nineties. The negative growth rate in the nineties indicated that PUBGCF is declining at an accelerated rate (though the level of significance is weak, at 15% level of significance) during this period. Thus, unlike at national level, GCFAGR in general and PUBGCF in particular start declining from nineties only.

This can be further endorsed when we look at the relationship between NSDP originating from agriculture (NSDPAR) and GCFAGR at 1980-81 prices.

TABLE 4.8

Trend of GCFAGR, NSDPAGR and AGGNSDP of Manipur

(Rs. in lakh)

Year	*GCFAGR*	*NSD-PAGR*	*col. 2 as per-cent of col. 3*	*AGGNSDP*	*col. 3 as per cent of col. 5*
1	*2*	*3*	*4*	*5*	*6*
1972-73	3.8057	5437	0.070	12181	44.635
1973-74	5.3899	7631	0.071	14775	51.648
1974-75	6.6854	8586	0.078	16290	52.707
1975-76	11.153	9111	0.122	17306	52.646
1976-77	12.970	8791	0.148	17492	50.257
1977-78	12.820	10347	0.124	19349	53.476
1978-79	16.493	9566	0.172	18743	51.038
1979-80	11.247	9193	0.122	18969	48.463
1980-81	11.791	9090	0.130	19956	45.550
1981-82	11.276	9286	0.121	21103	44.003
1982-83	12.185	8971	0.136	21445	41.833
1983-84	11.159	9557	0.117	23297	41.022
1984-85	12.664	9807	0.129	24257	40.430
1985-86	13.447	10131	0.133	25617	39.548
1986-87	12.621	9787	0.129	26125	37.462

contd...

1	*2*	*3*	*4*	*5*	*6*
1987-88	14.439	10167	0.142	28166	36.097
1988-89	14.374	9992	0.144	29526	33.841
1989-90	15.194	9583	0.159	29935	32.014
1990-91	19.615	10405	0.189	31622	32.904
1991-92	20.148	10774	0.187	34284	31.426
1992-93	13.847	10467	0.133	35910	29.148
1993-94	13.717	11485	0.119	36978	31.059
1994-95	14.308	12463	0.115	39644	31.437
1995-96	15.235	11708	0.130	40719	28.753
1996-97	16.291	12067	0.135	43406	27.800
1997-97	14.433	12832	0.112	46156	27.801

Source: 1. Same as Table 4.6

2. National Account Statistics of India: 1950-51 to 1996-97, EPW Research Foundation, 3rd Edition, Mumbai (India), Oct. 1998.

Note: 1. GCFAGR = Gross Capital Formation in Agricultural Sector of Manipur.

2. NSDPAGR = Net State Domestic Product originating from Agriculture.

3. AGGNSDP = Aggregate Net State Domestic Product of Manipur for the economy as a whole.

Table 4.8 revealed that later formed less than one percent of the former throughout the period. This shows that very little amount of income originating from agriculture in the state has been devoted to asset creation for further production of this sector. The NSDPAGR increased from Rs. 5437 lakh in 1972-73 to Rs. 9566 lakh in 1978-79 then declined to Rs. 8971 lakh in 1982-83 and remained stagnant till 1989-90. Thereafter it gradually increased to Rs. 12832 lakh in 1997-98 with some minor variation.

During the entire period under study NSDPAGR is growing at an annual compound growth rate of 1.86 percent. It grew at a decelerating rate of 6.55 and 1.05 percent in the seventies and eighties respectively and 2.96 percent in the sub-period of nineties. Though growing but low rate of capital formation in eighties seems to be rightly felt which is not as expected as in the nineties. In spite of negative growth rate of GCFAGR, NSDPAGR is growing at the rate 2.96 percent in the nineties. This may be mainly due to the favourable monsoon and diversification of cropping pattern.

Further, during the period under study i.e. 1972-97, the aggregate NSDP of Manipur (AGGNSDP) grew at the rate of 4.77 percent per annum. It grew at the rate of 5.9 percent in the seventies, 4.82 percent in the eighties and 5.22 percent in the nineties. In absolute term AGGNSDP has been steadily rising during the entire period from Rs. 12,181 lakh in 1972-73 to Rs. 46,156 lakh in 1997-98 with some minor fluctuation. When we compare the trend of AGGNSDP with NSDPAGR, the latter shows a wider range of variation than the former.

This is further endorsed by the annual compound growth rate of the two. The annual compound growth rate of AGGNSDP is higher than the NSDPAGR growth rate during the period 1972-97. Similarly, the decade-wise annual compound growth rate of AGGNSDP is higher than the NSDPAGR in the nineties and eighties but slightly lower in seventies. This implies that, though the agriculture sector continue to be the largest contributing sector to the states income, since the beginning of eighties there seems to be a sweeping change in the economic activities of the state in favour of non-agricultural sector.

4.6 Agricultural Capital Formation in Manipur Vs. All India

Having constructed and analysed the time series of agricultural capital formation in Manipur, let us now examine the comparative picture with respect to all India.

While comparing the agricultural capital formation in Manipur with all India figure we have to concentrate mainly on the time series of PUBGCF and to some extent GCFAGR only. In the case of PVTGCF, the two series are not comparable at all. This is because of the paucity of data and methodology being adopted for the estimation of PVTGCF in Manipur in the present study. Henceforth let us first look at the structure of PUBGCF in India and Manipur. In both the cases capital formation in irrigation related activities (major, medium and minor irrigation and flood control projects) constituted the largest chunk and its proportion has been growing over the year. At present over 90 percent of the PUBGCF came from irrigation related activities in both the cases.

When we compare the trend of PUBGCF, it is found that PUBGCF at all India increased till 1980-81, then declined during the entire decade of eighties but picked up in the nineties. While in the case of Manipur it increased till 1978-79 then declined suddenly in 1979-80 and remained stagnant till mid eighties then picked up slowly till 1991-92. Thereafter it declined. Thus unlike at national level, PUBGCF in Manipur shows a declining trend during the decade of nineties only. This can be further endorsed when we look at the annual compound growth rate given in Table 4.2 and 4.7. There is a high positive annual compound growth rate both in India and Manipur during the decade of seventies but it was exceptionally higher in the case of Manipur than all India. During the decade of eighties a negative growth rate (though insignificant) could be seen in the case of all India while there was positive growth rate in Manipur. But during the nineties the opposite was true. Thus, unlike at national level, PUBGCF in Manipur started declining only from the beginning of nineties. Similar inference could be drawn when we look at the trend of GCFAGR in India and Manipur.

Figure 4.4

Annual Growth Rate of Gross Capital Formation in Agriculture Sector of Manipur and India during 1973 to 1996 (at 1980-81 prices)

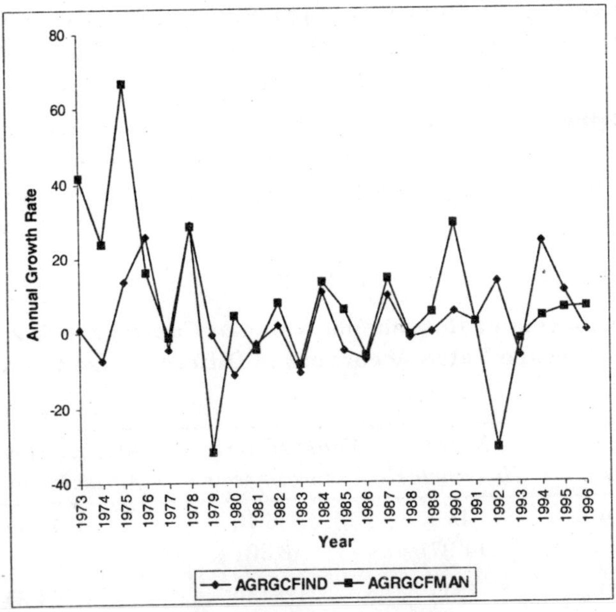

Next let us look at the year-to-year growth rate of GCFAGR in India and Manipur as shown in Figure 4.4. It is found that the direction of movement of annual growth rate of GCFAGR both in India and Manipur was the same till 1991; thereafter the two seems to be heading towards diverse direction. But during the period 1972-91, the magnitude of positive growth rate in Manipur is higher than the national level whereas in the case of negative growth rate the two moves at more or less in equal magnitude. That is to say that the year-to-year variation in agricultural capital formation in the case of Manipur is wider than all India, giving a picture of instability.

4.7 Analysis of Field Survey Data

As stated earlier the survey conducted in the year 2000 could not extract the data on private capital formation. Instead we could only arrive at the stocks of private capital. An analysis on the stock of private capital extracted from survey data reveals that the farm investments on fixed capital is an increasing function of the farm size. The distribution of households based on different size class of operational holdings[16] measured in acre is shown in Table 4.9 along with the value of agricultural assets owned by them.

TABLE 4.9

Distribution of Households, Value of Capital Stock and its Average Value According to Different Size Class

(in Rs.'000)

Size Class (in acre)	*No. of Households*	*Value of Agricultural Assets*	*Average Household Agricultural Assets*
0.1-1.0	44 (14.97)	636.70 (6.30)	14.47
1.1-2.0	131 (44.56)	3184.98 (31.53)	24.31
2.1-4.0	99 (33.67)	4198.09 (41.56)	42.40
4.1-8.0	16 (5.44)	1339.17 (13.26)	83.70
8.1 & above	4 (1.36)	743.04 (7.35)	185.76
Total:	294	10101.98	34.36

Sources: Field Survey.

Note: Figures in the parenthesis indicates percentage of the column total.

Table 4.9 shows that the average size of holdings in the valley area is 2.15 acres which is less than one hectare, while the average stock of capital owned by each cultivator household is Rs.34360. Out of the total households covered in the survey 15 percent of them operates on less than one acre and own only 6 percent of the total stocks of capital assets. The average value of stock owned by them is Rs.14,470 per cultivator. Further, about 45 percent of the households operate on one to two acres of land. They own only about 32 per cent of the total stocks of capital. On average each cultivator household belonging to this size class owns assets worth Rs. 24,310 per cultivator household. On the other hand, only about 7 percent of the households operating on four acres and above owned over 20 percent of the total capital stock which implies that on average every household of this size class owns capital assets worth Rs.1,04,110. That is to say that 60 percent of the cultivator households operate on holdings less than two acres of land. But they owned only about 37 percent of the total capital stocks. While 40 percent of the cultivator households operating on land more than two acres owned more than 63 percent of the capital stock. It can also be seen that the average value of capital stock owned by different size classes increases with the increase in the sizes of operational holdings.

Let us now look at the percentage distribution of capital stock in different size class of the four districts shown in Table 4.10. The percentage share of capital stock owned by cultivator households operating on less than two acres is almost similar in three districts viz. Imphal East, Imphal West and Bishnupur districts. However, in the Thoubal district the corresponding figure is much lower which is less than 20 percent. Similar is the case with size class two to four acres. On the other hand, the percentage share of capital stock owned by cultivator households with above four acres is similar in the three districts. But in the Thoubal district it is over 50 percent higher than rest of the districts. This implies that unlike in other districts fixed investments on farmers account in Thoubal district is much higher with larger holdings.

TABLE 4.10

Percentage Distribution of Capital Stock in Different Size Class in the Four Valley Districts of Manipur

Size Class (in acre)	*Imphal East*	*Imphal West*	*Bishnupur*	*Thoubal*
0.1 - 1.0	2.45	8.14	14.79	4.40
1.1 - 2.0	42.00	34.84	29.54	14.53
2.1 - 4.0	45.37	45.70	45.70	29.39
4.1 - 8.0	7.37	11.32	5.84	29.01
8.1 & above	2.81	-	4.13	22.67
Total:	100	100	100	100

Source: Field Survey.

The farm equipments used in the four districts are mostly of traditional types that can be formed with the application of unpaid family labour. In other words, most of the farm equipments used by the cultivators of the valley areas have been created without much of the financial saving. The number of cultivator households who own modern farm equipments such as tractors and power tillers are very few in members. Similarly, households with modern irrigation assets are also very few. This is mainly attributed to mono-crop (rice) system of cultivation. At the same time cropping intensity is also very low. That is to say that during the rabi season majority of the field are remained unused.

Let us now see the nature of composition of capital stock in the valley areas of the state shown in Table 4.11.

The total capital stock is composed of four major items. They are:

(a) Farm equipments,

(b) Live stocks,

(c) Farm house, animal sheds and golas, and

(d) Irrigation assets.

TABLE 4.11

Composition of Stock of Agricultural Assets by Different Size Class

(Rs. in '000)

Size Class (in acre)	*Farm Equipments*	*Live-stocks*	*Farm Houses & Animal Shed*	*Irrigation Assets*	*Total*
0.1-1.0	175.15 (27.51)	169.1 (26.56)	292.45 (45.93)	-	636.70 (100)
1.1-2.0	1000.70 (31.42)	853.66 (26.80)	1289.32 (40.48)	41.3 (1.30)	3184.98 (100)
2.1-4.0	1156.02 (27.54)	1196.82 (28.51)	1770.35 (42.17)	74.9 (1.78)	4198.09 (100)
4.1-8.0	640.57 (47.83)	265.00 (19.79)	407.60 (30.44)	26.00 (1.94)	1339.17 (100)
8.1 & above	480.34 (64.65)	58.7 (7.90)	169.00 (22.74)	35.00 (4.71)	743.04 (100)
Total	3452.78 (34.18)	2543.28 (25.18)	3928.72 (38.89)	177.2 (1.75)	10101.98 (100)

Source: Field Survey

Note: Figures in the parenthesis indicates percentage share of row total.

Table 4.11 shows that farm houses, animal sheds and golas constitute the largest share which accounts for about 39 percent of the total capital stock. They are closely followed by farm equipments with little over 34 percent. The share of irrigation assets is the least with only 1.75

percent. Further, an examination into the structure of asset composition in different size class reveals that the share of livestocks and farmhouses, animal sheds and golas declined with the increase in the size of holdings. The share of farm equipments and irrigation assets increases with the increase in the size of holdings. Thus, table no. 4.11 revealed that modern farm equipments and irrigation assets are mainly owned by the large size holdings. While small holders mainly depend on either bovine power or hired machinery and equipments from those who own modern farm equipments.

Next we shall look at the comparative picture of the percentage composition of agricultural capital stock in the four districts which is shown in Table 4.12.

TABLE 4.12

Percentage Composition of Agricultural Capital Stock in the Four Valley Districts of Manipur

Capital Stock	*Imphal East*	*Imphal West*	*Bishnupur*	*Thoubal*
Farm Equipments	29.93	45.01	21.71	40.33
Livestocks	25.53	22.88	39.95	15.97
Farm houses & Animal Sheds	41.67	30.35	38.00	42.59
Irrigation Assets	2.87	1.176	0.34	1.11
Total	100	100	100	100

Source: Field Survey.

The percentage share of farm equipments is highest in Imphal West district with 45 percent, which is followed, by Thoubal district with little over 40 percent. On the other hand, the percentage share of livestocks is highest in Bishnupur district where the percentage of farm equipments is the least. The percentage share of livestocks is the least in Thoubal district which constitutes about 16 percent. This is to say that farm equipments and bovine power are interchangeably used. In other words, the increase in the mechanization of farm would lead to decline in the use of bovine power. The percentage share of farm houses, animal sheds and golas are almost similar except in Imphal West district where the figure is slightly lower (30.35 percent). While the share of irrigation assets is very low in all the four districts. As stated earlier, the low percentage share of irrigation assets in all the four districts indicates that farmers in the valley areas of the state are mainly depend on the highly uncertain monsoon, mono crop system of cultivation and low cropping intensity. That is the entire farm sector cultivates only rice crop while in other seasons almost the entire fields remain unused.

4.8 Conclusion

In this chapter we have examined the structure and quantum of capital formation in agricultural sector of Manipur based on the methodology discussed in Chapter III. The following findings were emerged:

1. Throughout the period of study irrigation and flood control accounted for the major chunk of the total public capital formation in agriculture sector of Manipur.
2. Private capital formation per rural household has increased from 41 percent of the gross capital expenditure in 1971-72 to 93 percent of the gross capital expenditure in 1991-92. However the structure of private capital formation has completely changed in favour of other expenditure during the corresponding period.
3. Inspite of the low rate of capital formation in the state, public capital formation in particular and gross capital formation in general decline, more particularly in the nineties.

4. An analysis of the field survey data revealed that the farm investments on fixed capital assets is an increasing function of the farm size. That is to say that modern farm equipments are mainly owned by the more affluent farmers with larger farm land.

NOTES

1. For more detail see Shultz, T.W. (1964): *Transforming Traditional Agriculture*.
2. Mishra, S.N. (1996): "Capital Formation and Accumulation of Indian Agriculture since Independence," *Indian Journal of Agricultural Economics*, vol. 51, nos. 1&12, Jan - June, p. 193.
3. Shetty, S.L. (1990): "Investment in Agriculture: Brief Review of Recent Trends," *Economic and Political Weekly*, vol. XXV, nos. 7 & 8, pp. 393-398.
4. Mallick, S.K. (1993): "Capital Formation in Indian Agriculture: Recent Trends", *Indian Journal of Agricultural Economics*, vol. 48, no. 4, Oct.-Dec., p. 672.
5. *ibid.*, p. 673.
6. Gandhi, Vasant P. (1996): "Investment Behaviour in Indian Agriculture," *Indian Journal of Agricultural Economics*, vol. 51, no. 4, Oct.-Dec., p. 555.
7. Mitra, Ashok K. (1997): "Public and Private Investments in Agriculture," in Desai, Bhupat, M. (eds.) *Agricultural Development Paradigm for the Ninth Plan under New Economic Environment*, CMA Monograph, No. 179, Oxford & IBH Publishing Company, New Delhi, p. 967.
8. Kumar, A. Ganesh (1992): "Falling Agricultural Investment and its Consequences", *Economic and Political Weekly*, vol. 27, no. 42, Oct. 17, p. 2307.
9. Mishra, S.N. (1996): "Capital Formation and Accumulation in Indian Agriculture since Independence", *Indian Journal of Agricultural Economics*, vol. 51, nos. 1 & 2, Jan.-June, p. 201.

10. Chand, R. (2000): *Emerging Trends and Regional Variation in Agricultural Investments and their Implication for Growth and Equity*, National Centre for Agricultural Economics and Policy Research, New Delhi, p. 7.
11. Dantwala, M.L. (1986): "Strategy on Agricultural Development since Independence," in Dantwala, M.L. & Others (eds.): *Indian Agricultural Development since Independence: A Collection of Essays*, Oxford & IBH Publishing Co. Pvt. Ltd., New Delhi, p. 3.
12. The heads of capital expenditure are crop husbandry, diary development, fisheries, forestry and wildlife, plantations, food storage, warehousing, agricultural research and education, investment in agricultural financial institutions, corporations, other rural development programmes, special area programmes, hills areas, North-Eastern areas, major and medium irrigations, command area development, rural electrification, fertilizers industries and district roads.
13. Chand, R. (2000): *op. cit.*, p. 61.
14. These are soil and water conservation, crop husbandry animal husbandry, food storage, warehousing, dairy development, fisheries, forestry and wildlife, corporation, agricultural research and education, investment in financial institutions and other investments in agricultural programmes.
15. Gulati, A. and Seema Bathla (2000): *Capital Formation in Indian Agriculture: Trends, Composition and Implications for Growth*, (Mimeo), Institute of Economic Growth, Delhi, p. 70.
16. By operational holding we meant the total area in which the rice crop is cultivated mainly in the kharif season. It excludes the areas where residential buildings were located.

5

Determinants of Capital Formation in Agricultural Sector of Manipur

5.1 Introduction

Having constructed the time series of capital formation in agricultural sector of Manipur, a vital issue required to be analysed thoroughly is the identification of factors that determine agricultural capital formation. A proper identification of factors having direct bearing on agricultural capital formation is highly pertinent if we are to acquire capital-intensive modern agriculture. This would enable us to formulate proper strategy to combat any shortcoming in transforming traditional agriculture into modern agriculture.

There are a number of studies related to the issue under consideration using macro level aggregate data in India. However, there are a few studies at the regional level. The development of agricultural infrastructure with the investment from public account in different states/ regions within the states is highly unequal. Indeed, there are many rain fed/dry-farming areas operated with traditional tools and machinery in the country. Similarly, the flow of institutional credit and adaptability of new technology are also different accordingly. To that extent the determinants of capital formation in agriculture may also vary. In other words, factors that determine the capital formation in modern agriculture may not be the same as that of tradition-ridden agriculture. It is, therefore, highly pertinent to examine the factors bearing on agricultural capital formation at national level and at state/regional level.

In this chapter we shall attempt to identify the factors that determine capital formation in agriculture sector of Manipur. The study may, however, be limited to only few variables due to the lack of adequate database. First we shall have a brief survey of the literature on the issue under consideration.

5.2 Review of Literature

A survey of literature on the determinants of capital formation in Indian agriculture revealed that in macro modelling the determinants of private capital formation are fairly well identified in a behaviouristic framework while public capital formation are deemed to be exogenously determined. In fact, with the exception of few models public capital formation is assumed to be exogenously determined as it is considered to be dependent upon policy priorities and availability of fund in government account.

In recent years, deviating away from this general perception Dhawan and Yadav (1997) have attempted to establish a functional relationship for public investment in Indian agriculture. They considered savings in the public sector i.e. surplus on revenue account, net market borrowings and agriculture's share in total public capital formation in the economy as a whole as explanatory variables. Their study confirmed that, though all the three variables have positive effect, net borrowings by the government emerged as the most important factor that determined public capital formation in Indian agriculture. Each additional rupee of net borrowing by the government resulted in about forty paise worth of fixed capital formation in agriculture on public account over the period 1960-92. During this corresponding period, one rupee of additional savings in government's revenue budgets gives rise to about thirty paise worth of public capital formation. The contributory role of agriculture's share in total public capital formation emerges somewhere in between the other two variables.[1]

There are a number of factors that determine private capital formation in Indian agriculture. These factors are broadly classified into two categories—price and non-price factors. The price factors include all the types of farm outputs and inputs prices including wage rate and interest rate on borrowed funds. However, for the purpose of aggregate investment analysis, all these price factors are subsumed into a single factor called the 'terms of trade'. It is generally perceived that favourable terms of trade positively influence the farmers to invest more on their farm land. It, therefore, is expected to have positive impact on private capital formation in agriculture.

The terms of trade is not possible to relate with private capital formation in a cross sectional analysis. But in a country like India where inter state disparity in private capital formation is clearly visible from various decennial All India Debt and Investment Surveys, some researchers have attempted to examine the role of price factors in influencing capital formation on private account. Dhawan and Yadav (1995), thus, attempted to examine the relationship between private investments in 17 major states of India in 1981-82 with a price variable i.e. interest rate. It is a well known fact that a high rate of interest can discourage private investment. Though they found a negative relationship between private investment in agriculture and interest rate on term loan finance from different financial institutions but it is very weak.[2] Their conclusion reveals that some of the price factors are responsible for regional disparities in private investment in agriculture.

The non-price factors that determine the private capital formation in Indian agriculture are numerous. Dhawan and Yadav (1995) classified all those non-price factors into six sub-categories or groups. They are 1. Technology, 2. Institutions, 3. Government investments in infrastructure, 4. Government investment in agricultural sector, 5. Quantum and quality of government services having a direct bearing on farm products and marketing, and 6. Other factors.[3] A number of empirical studies showing why and how each of the factors placed in particular sub-categories acquired special importance in the growth of modern farm as well as non farm sector and hence private fixed farm investment can be obtained from Dantwala and other (1996), Ratha and Sarma (1992), Desai (1997), Pal and Singh (1998), Karmakar (1998), W.K. Singh (1999), Mohanty (1994), Chattopadhyay and Atamu Sengupta (1997) Bhatia (1999), etc.

In recent years some researchers felt that there exists some linkage between price and non-price factors, which might play some crucial role in explaining the variation in private investment over the years. Misra and Hazell (1997) have attempted to show the importance of interactive role of high yielding varieties (HYV) technology and terms of trade. They concluded that, in modernized phase i.e. during 1981-82 to 1993-94, the interactive term between terms of trade and technology becomes the most important factor in explaining the variation in private investment.[4]

Several studies have concluded that public investment has emerged as an important determinant of private investment whether termed as complementarity effect or inducement effect, the implication being that private investment grows, or falls with public investment. However in the wake of divergent movement of public and private investment starting from late seventies some researchers have challenged the earlier conclusion. Prior to the discussion of other determining factors we shall first see the detailed arguments on the relation between public and private capital formation in Indian agriculture.

There are two views in this regard, in the literature of capital formation in Indian agriculture. First, their exists/ does not exist a complementarity between public and private capital formation. Second, public capital formation has an inducement effect on private capital formation. However, the complementarity word is often used in a broader sense to include inducement effect as well. Thus, by complementarity, it has generally been understood that capital formation on government account stimulates private capital formation. In other words, the positive association between public and private capital formation has been interpreted as complementarity effect. Similarly, in macro modelling parlance, some researchers described such stimulation as 'crowding in' effect while the phenomenon of hindering private investments as 'crowding out' effect.[5]

The complementarity between public and private capital formation in Indian agriculture is often examined both on the basis of the direction of movement and the statistical relationship between the two series. A recapitulation of the CSO series revealed that both public and private capital formation moved in the same direction till 1985-86, albeit weakened somewhat since late seventies, indicating a strong positive association between the two series. Such positive association between the two series has generally been interpreted as having complementarity effect. However, since 1986-87, the two series were heading in diverse directions and the association between the two series turned out to be negative. This has generated an intense debate in the country regarding the complementarity issue between public and private capital formation in Indian agriculture.

A survey of literature on the relationship between public and private capital formation in Indian agriculture revealed a widespread acceptance of the hypothesis that public capital formation provides a significant stimulus to private capital formation thereby serving as a powerful engine of stabilization. In other words, studies of different researchers based on CSO series concluded that there exists a strong positive association between the two series in Indian agriculture (Shetty, 1990, Mallick, 1993, Wagle, 1994, etc) and later by Dhawan, (1996a, 1996b and 1998), Alagh (1997), Singh (1997), Karmakar (1998), etc.

While considering the stimulating role of public capital formation on private capital formation, Shetty (1990) concluded that the productive private investment in agriculture is largely influenced by public investment in agriculture.[6] Similarly, Mallick (1993) concluded that, "To a large extent, public and private investments are complementary, rather than substitutes for each other, and thus falling public investment may be affecting private capital formation. Hence, it is the increase in public investment in agriculture which influences the trends in private investment in agriculture."[7]

Further, according to the estimate of wagle (1994), one percent rise in government investment strengthens private investment in agriculture by 0.43 percent in the short run and 0.87 percent in the long run.[8] The claim of the existence of complementarity between the two series is well supported by the study on the series of agricultural investment data for the period of 1952-80 constructed by Gandhi (1990) based on the economic theory of investment behaviour. He concluded that, "... government investment in agriculture significantly stimulates private investment, thereby having a multiplier effect on raising capital stock available for productive activities in the rural areas. This increase in capital stock, apart from directly contributing to growth, can also help in removing constraints to rapid technological change. Furthermore, since some capital stock itself embodies new technology, it may make a direct contribution to technological change and higher productivity in agriculture."[9]

For the first time Mishra and Chand (1995) have strongly refuted the earlier claims of complementarity between public and private capital formation and questioned the validity of any relationship between the two on the basis of conceptual, methodological and factual issues. According to them there is no apparent complementarity between public and private investment in Indian agriculture. Private investment may be partly induced by public investment or partly autonomous. The effect of public investment on private investment is refuted on the ground that complementarity should be interpreted as a kind of relationship to form unity, which in turn produce an output. Neither of the complement part can produce all by itself. Further, they also opined that complementarity is not a causal relationship like that of inducement effect and hence can not be studied with the help of statistical relationship.[10] They were well supported by Mitra (1997), Misra and Hazell (1996), Misra (1998) and Chand (2000). According to Mitra (1997) complementarity is applicable only at micro level, which cannot be generalized at macro level. He concluded that, "Given the factors determining the investment behaviour of the farmers, the private investment in agriculture may be partly induced by public investment in agriculture, which may be more significant, and partly be autonomous, some of which may have been influenced by aggregate public investment on infrastructure not specific to agriculture sector."[11]

The complementarity hypothesis was considered so obvious that its refutation has generated a critical review of the investment literature of this sector. A number of researchers have attempted to re-examine the issue. Dhawan and Yadav (1995) were the first to work in this direction using the data from AIDIS 1981-82, RBI to prepare the estimates of per cultivator fixed capital formation in agriculture on private account for 17 major states of India in 1981-82. They analysed its relationship with investment in canal irrigation in cumulative terms as proxy variable for public investment along with other relevant variables. The study found that practically in all the states there was a high correlation (r=0.65) between the two variables i.e. per cultivator fixed capital formation during 1981-82 and investment in canal irrigation in cumulative term by 1981-82.

Since public investment in irrigation constitutes a lion's share in total public investment in agriculture in India reduction in their magnitude could explain the deceleration in FCFA on private account during the 1980s—the annual rate of increase in FCFA on private account dipped from the sizable level of 4 percent during the 1970s to merely 0.4 percent during 1980-88. Thus, they concluded that complementarity between public and private investments in agriculture cannot be given up in favour of the substitutability thesis being suggested on the basis of recent data from the CSO series.[12]

Further, Dhawan (1996a) opined that negative relationship (elasticity = -0.50) between public and private investments obtained by Mishra and Chand (1995) for the period 1980-81 to 1991-92 could be attributed to their failure to explore the relationship in multiple regression framework and restriction of the analysis mainly to the period of eight years when public investment declined substantially while private investments continued to rise.[13] Similarly, Alagh (1997) opposed the conclusion drawn by Mishra and Hazell (1996) on the basis of simple regression between public and private investments for the period of 1980-81 to 1989-90 only. He is also critical of the data used by them and insists on using gross fixed capital formation instead of gross capital formation since the later includes change in stock. He concluded that, "Since the effects of changes in public investments can take time to impact on induced private investments, a simple regression of private agricultural gross capital formation on public agricultural gross capital formation for nine years of data showing a weakly significant relation is no refutation of 'crowding in' hypothesis."[14] Gandhi and Mani (1997) echoed the lack of degree of freedom while analysing the relationship for the period of 1980-92. They observed different behavioral patterns with negative and significant coefficient of government capital stock representing government investment. Thus, they concluded that, "a fall in the government investment in the last decade has been accompanied by a compensatory rise in the private investment in the recent years. This structural change would give a negative sign-and this does not indicate a negative impact of government investment."[15]

While considering the stimulating role of public capital formation on private capital formation, Dhawan (1996a) observed that public capital formation in canal irrigation has positive bearing on the private capital formation. The elasticity of private fixed capital formation in agriculture with respect to canal irrigation is 0.25 during 1981-82.[16] Thus, he reasserted the prevalence of complementarity between public and private capital formation both at macro and micro level in Indian agriculture. To quote him, "Since development of canal irrigation accounts for a major share of total public investments in Indian agriculture, the veracity of the proposition gives credence to the complementarity hypothesis according to which private fixed capital formation in Indian agriculture is generally positively induced by corresponding fixed capital formation on public account."[17]

The methodology adopted by Dhawan, net canal irrigated area as percentage of net sown area instead of public investment as the relevant explanatory variable in the fixed capital formation at the household level, however, came under severe criticism.

The debate shows a serious disagreement on whether complementarity between public and private investments in Indian agriculture still persists. Chand (2000) once again questioned the complementarity at the macro level by re-defining and re-estimating public investment in Indian agriculture using capital outlay data on 23 major heads of expenditure[18] related to agriculture given in finance account. The study re-examined the complementarity relationship by using statistical relationship of multiple regression framework. The results showed a negative but non-significant impact of public investment defined under broad series on private investment. Thus, he concluded that, this was indicative of a lack of complementarity between public and private investments at the aggregate level. However, contrary to this, the analysis across the different states of India revealed that the estimated broad series bears a positive and significant impact on private investment in 1981-82 and positive but insignificant impact in 1991-92. Therefore he concluded that only at different regional settings partial complementarity on the types of public investment could not be ruled out. But contrary to his empirical result at macro level, he repeatedly asserted the role and importance of public investment in creating infrastructure and promoting long-term agricultural growth.[19]

The conclusion drawn by Chand (2000) has encouraged some researchers to dig deeper and visualise the nature of relationship between the two types of investments from the very concept and coverage and re-estimating the public sector investment in Indian agriculture (Gulati and Seema Bathla, 2000). They attempted to test the complementarity relationship using both physical and financial terms of public sector investment of CSO series as well as broad series[20] defined by them. Unlike the other studies they used the cumulative investment approach (both physical and financial) for establishing statistical relationship using multiple regression framework. The study concluded that, "complementarity relation between public and private investment in agriculture can be established with the official data on investment in major and medium irrigation works as well as with new estimates based on sound conceptual grounds, which ever way they are defined."[21]

Thus, one can conclude that the complementarity relationship between public and private capital in Indian agriculture, particularly during '80s and '90s, is still yet to be resolved. But majority of them are in favour of complementarity between the two series. The debate has been quite useful in understanding the complex nature of relationship between the public and private capital formation in particular and overall capital formation in agriculture in general.

Apart from public capital formation, different authors have identified different sets of factors as determinants of private capital formation in Indian agriculture. The most significant factors that have emerged are terms of trade, high yielding variety (HYV) technology, institutional credit (both cooperative and commercial banks), rural savings and wage rates. The interactive role of HYV technology and the terms of trade were also highlighted by some researchers. However, the relative importance of these factors varied depending upon the period under study, coverage and methodology being adopted by different researchers.

The study on private investment behaviour by Gandhi (1990) for the period 1952-80 identified a set of factors determining private investment in Indian agriculture. The determinants in order of importance as indicated by the elasticity of private investment with respect to each one of them are as follows: the stock of private capital in agriculture in the preceding year (-5.29) cooperative credit (+0.89), value of agricultural output (+0.58), rural savings (+0.56), the stock of public capital in agriculture in the same year (+0.42), agricultural output price (+0.31), commercial bank credit (+0.31), and agricultural wage rate (+0.30).[22] But when the period of study is extended to post 1980s i.e. till 1992-93, when public investment declined but private investment with variation increased, the relative importance of these factors have changed. The later study found that rural savings and cooperative credit to agriculture emerged as the strongest determinants. They were followed by HYV coverage, agricultural wages and commercial bank credit.[23] Similarly, Misra and Hazell (1997) showed the relative importance of interaction term between the terms of trade and technology in modernized phase (i.e. 1981-82 to 1993-94) which was not significant in the initial phase (1966-67 to 1980-81).[24]

On the other hand, when the analysis is restricted to only CSO series for the period of 1961-62 to 1995-96, Misra (1998) shows the relative importance of terms of trade and technology measured as the percentage of area under high yielding varieties of seeds.[25] But according to the studies conducted by Chand (2000) and Gulati and Seema Bathla (2000), terms of trade and institutional credit advanced to farmers have significant positive impact on private investment in Indian agriculture.

Having gone through a brief literature survey on the issue of determinants of private capital formation and its complementarity relationship with public capital formation in Indian agriculture, an attempt would be made to critically appraise the same at the regional level. The basic questions are: what are the determinants of private capital formation in agriculture sector of Manipur? Does the complementarity relationship between public and private capital formation in agriculture sector of Manipur exist or not?

5.3 Determinants of Agricultural Capital Formation in Manipur

As in the national level, public capital formation in agriculture sector of Manipur has been treated as exogenous variable. This is because public capital formation in agriculture depends upon the policy programme and availability of fund on the part of the government of Manipur. The present study shall, therefore, concentrate mainly on the determinants of private capital formation in agriculture sector of Manipur using behaviouristic framework of both simple and multiple regression. However, the analysis may be limited to only few variables due to the lack of adequate database.

5.4 Specification of Variables

Private capital formation in agriculture sector of Manipur is taken as a function of public capital formation lagged by one year, terms of trade, technology and interaction between terms of trade and technology. The functional form that can be used is;

$$PVTGCF_t = f(PUBGCF_{t-1}, TT_t, TECH_t, INT_t)$$

where,

$PVTGCF_t$	=	Private capital formation at 1980-81 prices (Rs.)
$PUBGCF_{t-1}$	=	Public capital formation at 1980-81 prices (Rs.) lagged by one year.
TT_t	=	Index of terms of trade.
$TECH_t$	=	Technology
INT_t	=	Interaction term between terms of trade and technology.

The expected relationships of these independent variables with private capital formation in agriculture are stated below.

5.4.1 Public Capital Formation ($PUBGCF_{t-1}$)

The activities in which public capital formation in agricultural sector generally flows in and the impact and externalities generated by those activities, which would facilitate the individual farmers to invest more, are now well recognized. The survey of literature on the relation between the public and private capital formation given in the previous section indicated that the public capital formations strongly influence the farmers' decision to invest in their farms. Some scholars have strongly objected to this claim in the nineties when public capital formation declined persistently starting from mid '80s. As it has been done at the national level, the present study would also attempt to examine the effect of public capital formation on the farmers' decision to invest in the agriculture sector of Manipur. Hence, public capital formation at 1980-81 prices has been introduced in the equation and is expected to have positive impact on private capital formation in agriculture over the period 1972-73 to 1997-98.

Since capital formation on public account in the current year is not expected to influence the private investors to invest more in their farms in the current year itself, we have, therefore, used the PUBGCF lagged by one year for examining their impact on private capital formation in agriculture sector of Manipur.

5.4.2 Terms of Trade (TT_t)

The favourable terms of trade in agriculture can create stronger base for further growth in the agriculture sector as they are expected to increase profitability thereby inducing the farmers to invest in agriculture. The terms of trade, therefore, is expected to have positive impact on private capital formation over the period of time.

Three different ways of measurement of terms of trade for agriculture are available in the literature. They are First, Ratio of price index of agricultural products to price index of manufactured products as published in Economic Survey, Government of India. Second, Ratio of prices received by farmers for their products to prices paid by them for the commodities purchased from the non-agricultural sector and is called the barter terms of trade. It is prepared by the Commission for Agricultural Cost and Prices (CACP), Government of India. Third, Ratio of agricultural to non-agricultural GDP deflator called the gross terms of trade. This does not represent the prices at which quantities were traded among sector, what is measured is the relative valuation of agricultural and non-agricultural products. [26]

Though different scholars have used different concepts of measurement while examining its effect on private capital formation at national level, the implication remained the same. That is favourable terms of trade are envisaged to persuade the farmers to invest more in farm activities due to the enhancement of the overall returns to fixed investment in agriculture. It would not be possible to test the impact of terms of trade based on different measures for the present study due to the lack of adequate database; we shall, therefore, use gross terms of trade only to examine its impact on private capital formation.

Since farmers' expectations are generally based on two to three years' experience of price movements, three yearly moving averages of gross terms of trade has been considered for determining its impact on private capital formation in agriculture sector of Manipur.

5.4.3 Technology ($TECH_t$)

It is a well known fact that technology is an important factor moving the production frontier upward. Introduction of new technology to new areas would lead to better utilisation of available infrastructure and resources resulting into higher production thereby persuading the farmers to invest more.

By technology we mean the knowledge embedded in the inputs used and the change in technology may be defined as the change or shift in the production function.[27]

It manifests itself in the use of new inputs and knowledge leading to an upward shift of the production function in the long run.[28] A package of high yielding varieties (HYVs) of seeds, fertilizers, pesticides, insecticides, improved mechanical tools and equipments etc. together with assured water supply have been responsible for success of Green Revolution in Indian agriculture in mid sixties.[29] Thus, technology by itself in general involves all these aspects. However, it would not be possible to incorporate all these variables in the present study due to want of adequate database.

At the national level, technology has been introduced in the equation as the percentage of area under HYV seeds to total gross cropped area. But for the present study, due to the non-availability of complete time series on total gross cropped area and since rice is the only major crop in the state, we shall instead use the percentage of area under HYV of rice crop to the total area under rice crop as the proxy variable. As in the case of Indian agriculture, technology variable is expected to have a positive coefficient as higher the area under HYV higher would be the production thereby inducing the farmers to invest more.

5.4.4 *Interaction between Technology and Terms of Trade (INT_t)*

The increase in output with the introduction of new technology influences the farmers' decision to invest in agriculture in two ways acting in opposite direction. First, it would reduce the relative prices of the agricultural commodities depressing thereby terms of trade which in turn reduces the profitability of the farmers and hence the incentive for investment. Second, it would reduce the per unit cost of production leading to higher profitability and hence induce the farmers to invest more.[30] The interaction term between the terms of trade and technology is expected to capture the forces operating behind the two in the growth of private capital formation in agriculture. Thus, interaction term has been introduced in the equation for explaining the variations in private capital formation as has been done in many studies at national level (Misra and Hazell, 1996 &1997 and Misra, 1998).

5.5 Model specification

For analysing the factors influencing the private capital formation in agriculture sector of Manipur, the linear and log-linear simple and multiple regression equations are considered. As this has generally been done in various studies, selection of model is based on the criteria of theoretical relation involved, $\overline{R}^2$, sign and statistical significance of variables, the absence of both autocorrelation and heteroscedasticity, and the stability of the functional relationship.

$$PVTGCF_t = a + b_1\, PUBGCF_{t-1} + b -{}_2TT_t + b_3\, TECH_t + b_4\, INT_t + u_t$$

Where 'a' is the intercept term and b's are the coefficients of the independent variables. But one has to be very careful while interpreting the coefficient. This is because, if the equation is a simple linear one, then they are the simple coefficient of the respective variables. In the log-linear equation these coefficients represent the elasticity of private capital formation with respect to particular independent variable. u_t represent the error term.[31]

5.6 Findings

The degree of association between private capital formation in agriculture sector of Manipur and other relevant variables during the period 1973-74 to 1997-98 can be seen from the correlation coefficients presented in Table

5.1 (The raw data of terms of trade, technology and interaction term is given in Appendix 5.1).

TABLE 5.1

Correlation between PVTGCF and Other Relevant Variables

	$PVTGCF_t$	$PUBGCF_{t-1}$	TT_t	$TECH_t$	INT_t
$PVTGCF_t$	1.00	0.65005	0.57405	0.64451	0.66486
$PUBGCF_{t-1}$		1.00	0.26858	0.67679	0.63793
TT_t			1.00	0.66814	0.77015
$TECH_t$				1.00	0.97088
INT_t					1.00

Note: Absolute value of correlation coefficient above 0.505 is significant at 1% and the value above 0.396 is significant at 5% level of significance.

The results are for one year lag in public capital formation. The correlation coefficients between private capital formation and interaction term, one year lag public capital formation, technology and terms of trade are 0.665, 0.65, 0.645, and 0.574 respectively. This means that, the private capital formation in agriculture shows highly significant positive association with all the four variables under consideration.

High correlation coefficient may imply that private capital formation in agriculture sector of Manipur may be affected by all these four variables. In order to examine the effect of these variables on change in private capital formation in agriculture both linear and log-linear simple and multiple regression analysis was carried out. Let us first examine the results of the simple regression analysis.

When each of the four variables is used alone as explanatory variable to examine the impact on private capital formation, all the estimated coefficients are significant at one percent level of significance with expected sign. But all the specifications suffer from the problem of autocorrelation. The presence of autocorrelation may be mainly due to the methodology being adopted - massaging technique of interpolation and extrapolation of private capital formation series on the basis of some ad-hoc assumption that they would increase at compound growth rate.[32] The presence of autocorrelation affects the efficiency testing of the OLS. It, therefore, is highly pertinent to correct the problem with the help of some standard remedial measures. Here the problem has been corrected with the help of Cochrane-Orcutt iterative process.[33] The estimated linear regression coefficients along with their summary statistics, $\bar{R}^2$ and corrected value of autocorrelation using data for the period 1973-74 to 1997-98 are presented in Table 5.2.

TABLE 5.2

Regression of PVTGCF and Related Statistics

Independent Variable	*Estimated Regression Coefficient*	$\bar{R}^2$	*DW*
$PUBGCF_{t-1}$	0.015051 (1.447)	0.3975	1.6771
TT_t	(-) 0.0097997 (- 2.199)**	0.3004	1.764
$TECH_t$	0.0015986 (0.324)	0.39	1.524
INT_t	(-) 0.0000004 (- 0.012)	0.4178	1.5104

Note: 1. Figures in parenthesis under each estimated coefficients are the t-statistics.
2. Problem of Autocorrelation is corrected with the help of Cochrane-Orcutt iterative process.
3. ** are significant at 5% level of significance.

The Table indicates that only the terms of trade has significantly (at 5% level of significance) affected the private capital formation in agriculture. However, the negative sign on estimated coefficient of the terms of trade is contrary to the theoretical perception hence it needs to be closely scrutinized. Other remaining variable does not show any significant impact, though accompanied with expected sign of their impact on private capital formation.

When the variables are transformed in log form the level of significance of the estimated coefficients and value of $\overline{R}^2$, which are the main criteria of choosing a particular model, are higher than the simple linear regression analysis. The summary results of the log-linear regression analysis after the correction of autocorrelation problem are shown in Table 5.3.

TABLE 5.3

Regression of Log (PVTGCF) and Related Statistics

Sl. No.	*Independent Variable*	*Estimated Regression Coefficient*	$\overline{R}^2$	*DW*
1.	Log ($PUBGCF_{t-1}$)	0.091122 (1.955)***	0.4872	1.8819
2.	Log (TT_t)	(-) 0.61187 (-2.856)*	0.2704	1.9477
3.	Log ($TECH_t$)	0.054844 (1.019)	0.4896	1.5428
4.	Log (INT_t)	0.015309 (0.294)	0.5174	1.4478

Note: 1. Figures in parenthesis under each estimated coefficient are the t statistics.

2. Problem of Autocorrelation is corrected with the help of Cochrane - Orcutt iterative process.

3. * & *** are significant at 1% and 10% level of significance.

The results indicate that the estimated coefficient of one year lagged public capital formation is significant at 10 percent level of significance with expected positive sign. It explained nearly 49 percent, as compared to 40 percent in simple linear regression of the change in private capital formation in agriculture. The positive and significant estimated coefficient with high correlation coefficient (r = 0.65) means that the public capital formation in agriculture did stimulate private capital formation in agriculture though the magnitude of elasticity of private capital formation with respect to former is 0.09 only. In other words the complementarity hypothesis between public and private capital formation does prevail in the agriculture sector of Manipur as well.

Unlike the result before the correction of autocorrelation problem the estimated coefficient of terms of trade is found to be negative and significant at one percent level of significance. Though the value of $\overline{R}^2$ is small (0.27) the magnitude of elasticity of private capital formation with respect to terms of trade is very high (-0.61). The negative and significant estimated coefficient of the terms of trade implies that the favourable terms of trade retards the private capital formation in agriculture sector of Manipur. This is contrary to the theoretical perception that favourable terms of trade stimulate private capital formation in agriculture. The negative and significant impact of the terms of trade on private capital formation may be attributed by subsistence mode of farming and switching of farm activities to non-farm activities in Manipur. Otherwise the favourable terms of trade during the period 1973-74 to 1997-98 might have resulted into positive coefficient.

The equation number 3 and 4 of Table 5.3 indicate that the level of technology as measured by the percentage of area under HYVs of rice cropped to the total area under same crop and the interaction term between the terms of trade and technology do not significantly influence the change in private capital formation. This may be due to the use of proxy variable while examining the technological impact on private capital formation. However, we observe the positive sign on the estimated coefficient and relatively high $\overline{R}^2$ value (0.49 and 0.52 respectively). Thus, we can still conclude that they might influence the change in private capital formation in agriculture sector of Manipur. The elasticity of private capital formation with respect to the former is 0.055 while with respect to latter is 0.028.

Private capital formation in agriculture sector of Manipur may be simultaneously affected by all these four variables. In order to examine the simultaneous effect of these variables, multiple regression analysis was carried out. It would, however, be seen from Table 5.1 that different variables used as explanatory variables have highly significant correlation among them with the exception between one year lag public capital formation and terms of trade.

The correlation coefficient between one year lagged public capital formation and technology and one year lagged public capital formation with interaction terms are 0.677 and 0.638 respectively. Similarly, the correlation between terms of trade and technology, terms of trade and interaction term, and technology and interaction term are 0.668, 0.77, and 0.971 respectively. High correlation among different explanatory variables may cause the problem of multicollenearity, which affects the significance level of regression coefficients of those affected variables. The problem is more severe in the case of linear multiple regression analysis. Hence it becomes very difficult to estimate the precise impact of any explanatory variable on the dependent variable.

Transformations of variables in log form help in overcoming the problem of multicollinearity but the results still are not satisfactory. Nevertheless, we can have some idea about the nature of relation between dependent and explanatory variables by running regression of different combinations of explanatory variables in log form. Their results are shown in Table 5.4.

TABLE 5.4

Log Linear Multiple Regression between $PVTGCF_t$ and Different Combination of Explanatory Variables

Sl.No.	*Log($PUBGCF_{t-1}$)*	*Log(TT_t)*	*Log($TECH_t$)*	*Log(INT_t)*	$\overline{R}^2$	*DW*
1	0.12318 (2.157)**	-0.27457 (-0.913)			0.6603	1.6167†
2.	0.1152 (1.956)***		0.074544 (1.2)		0.5177	1.6471†
3.	0.13195 (2.222)**			0.038939 (0.656)	0.5570	1.6663†
4.		-0.59615 (-2.556)**	0.010085 (0.197)		0.513	1.9607
5.		-0.61943 (-2.778)**		-0.0084834 (-0.183)	0.5036	1.9271
6.	0.046268 (0.911)	-0.51305 (-2.042)***	-0.0014111 (-0.027)		0.6516	2.0845
7	0.051963 (1.021)	-0.5175 (-2.12)**		-0.10526 (-0.792)	0.6544	2.0593
8.	0.051266 (0.994)	-0.45631 (-1.738)***	0.072858 (0.694)	-0.078649 (-0.821)	0.6386	2.0465

Note: 1. Figures in parenthesis under each estimated coefficient are the t-statistics

2. † Problem of autocorrelation is corrected with the help of Prais-Winsten estimator while in remaining equations autocorrelation problem is corrected with the help of Cochrane-Orcutt iterative process.

3. *, ** & *** are significant at 1%, 5% and 10% level of significance

When all the four variables are considered, equation no. 8, only the terms of trade shows significant (at 10 percent level of significance) impact on private capital formation. But the sign of the estimated coefficient is negative which is similar to our earlier result. The other remaining three variables do not show any significant impact. However, the signs of the estimated coefficient of one year lagged public capital formation and technology are on the line of the theoretical perception. The negative sign of the estimated coefficient of the interaction term may be associated with the negative outcome of the favourable terms of trade.

When private capital formation is regressed on different combinations of explanatory variables, one year lagged public capital formation shows positive significant impact on private capital formation (equation number 1 to 3). While equation no. 4 to 8 revealed that the favourable terms of trade exerts negative and significant impact on private capital formation. However, the magnitude of elasticity of private capital formation with respect to terms of trade varies from 0.27 to 0.62 with the variation in the combination of explanatory variable. Similarly, the elasticity of private capital formation with respect to one year lagged public capital formation varies from 0.05 to 0.13. The impact of technology on private capital formation is inconclusive as the estimated coefficients are not significant.

As stated earlier the hypothesis of complementarity between public and private capital formation in agricultural sector of Manipur can not be rejected. We finally attempt to examine the complementarity issue between the two series with the help of estimated correlation coefficient between public and private capital formation at constant prices with different time lags of the former.

Further, the same has also been attempted to examine by expressing private sector capital formation in agriculture as a function of public sector capital formation in agriculture with different time lags. In other words, the two series have been used for the estimation of correlation and regression coefficients by assuming that private capital formation in agriculture follows the public capital formation with current year, one year, two years, and three years lag. The results of the estimated coefficients of correlation are presented in Table 5.5.

TABLE 5.5

Correlation between Private Capital Formation and Current and Lagged Public Capital Formation

PUBGCF Time Lag	$PVTGCF_t$
Current ($PUBGCF_t$)	0.62486*
One year ($PUBGCF_{t-1}$)	0.65005*
Two year ($PUBGCF_{t-2}$)	0.56594*
Three year ($PUBGCF_{t-3}$)	0.48396**

Note: * & ** significant at 1% and 5% level of significance respectively.

The Table shows that all the estimated coefficients are positive and significant. The value of correlation coefficient is highest with one year lagged public capital formation. Thereafter, its value has been declining with the increase in length of time lags. The positive association between the two series with different time lag implies that a rise in public capital formation would lead to rise in private capital formation in agriculture. However, the degree of association between the two series declined with increase in length of time lags.

TABLE 5.6

Log Linear Regression between (PVTGCF) and Different Lag Period of PUBGCF

Sl. No.	*Independent Variable*	*Estimated Regression Coefficient*	$\overline{R}^2$	*DW*
1.	Log ($PUBGCF_t$)	0.08999 (2.019)***	0.4939	1.729
2.	Log ($PUBGCF_{t-1}$)	0.091122 (1.9551)***	0.4872	1.8819
3.	Log ($PUBGCF_{t-2}$)	0.057167 (1.146)	0.3674	1.5286
4.	Log ($PUBGCF_{t-3}$)	(-) 0.035473 (-0.716)	0.2308	1.5806

Note: 1. Problem of Autocorrelation is corrected with the help of Cochrane-Orcutt iterative process.
2. Figures in Parenthesis under each estimated coefficient are the t-statistics.
3. *** significant at 10% level of significance.

Alternatively, the estimates of linear regression coefficients (log form) turn out to be positive till the two years lag. But the estimated coefficient is significant at 10 percent level of significance only for current year and one year lag, as shown in Table 5.6, indicating that one percent increase in public capital formation leads to about 0.09 percent increase in the private capital formation in both current year and one year lag public capital formation. Beyond one-year lag, the coefficients are insignificant. At the same time the explanatory power of public capital formation on change in private capital formation also declined from 0.49 percent to 0.23 percent with the increase in length of lag years. The complementarity between public and private capital formation prevailed in agricultural sector of Manipur.

5.7 Conclusion

In this chapter we have examined the determinants of private capital formation in agricultural sector of Manipur using simple and multiple regression framework. The following findings emerge:

1. The coefficients of correlation between different explanatory variables under consideration are very high. This has introduced the problem of multicollinearity in linear multiple regression analysis. In order to solve the problem multiple regression analysis in log form is considered while identifying the factors determining the private capital formation in agriculture sector of Manipur
2. The private capital formation in agriculture sector of Manipur is positively determined by public capital formation. However, the favourable terms of trade on agriculture exerts adverse impact on private capital formation, which is contrary to our theoretical perception. This may be attributed to the subsistence nature of farming and switching of farm activities to non-farm activities in Manipur.
3. The positive and significant sign on estimated regression coefficient of public capital formation implies that complementarity between public and private capital formation in agriculture sector of Manipur cannot be ruled out. The stimulating effect of public capital formation on private capital formation would carry on upto one year lag of public capital formation.
4. The elasticity of private capital formation with respect to public capital formation in current year and with one-year lag is almost similar (0.09). But with two years lag it declined to 0.057.

5. No definite conclusion can be drawn in the case of interaction term between terms of trade and technology. This may be attributed to the weak response of the technological factor and negative impact of the favourable terms of trade.

NOTES

1. Dhawan and Yadav (1997) "Public Investment in Indian Agriculture: Trends and Determinants", *Economic and Political weekly*, vol. XXXII, no. 14, April 5, pp. 710-714.
2. Dhawan and Yadav (1995): "Private Fixed Capital Formation: Some Aspects of Indian Farmers' Investment Behaviour", *Economic and Political Weekly*, vol. XXX, no. 39, Sept, p. A-107.
3. For more detailed analysis see Dhawan, B.D. (1996): "Trends and Determinants of Capital Investments in Agriculture", *Indian Journal of Agricultural Economics*, vol. 51, no. 4, Oct-Dec, pp. 529-542.
4. Misra and Hazell (1997): "Price and Non-Price Factors in Agricultural Investments", *Economic and Political Weekly*, vol. XXXII, no. 31, August 2, p. 1992.
5. For a detailed analysis on 'crowding in' and 'crowding out' effect of public investment see Sundararajan and Thakur (1980): "Public Investment, Crowding out and Growth: A Dynamic Model Applied to India and Korea", *International Monetary Fund Staff Papers*, December, pp. 814-859.
6. Shetty, S.L. (1990): *op. cit.*, p. 397.
7. Mallick, Sushanta Kumar (1993): "Capital Formation in Indian Agriculture: Recent Trends," *Indian Journal of Agricultural Economics*, vol. 48, no. 4, Oct.-Dec., pp. 675-676.
8. Wagle, M.P. (1994): "Estimates of Aggregate Functions for Demand for Fertilisers and Private Investment in Indian Agriculture," *Indian Journal of Agricultural Economics*, vol. 49, no. 1, Jan.-March, p. 63.
9. Gandhi, Vasant P. (1990): "Investment Behaviour in Developing Countries: The Case of Agriculture in India", *Food Research Institute Studies*, vol. XXII, no. 1, Washington D.C., p. 65.

10. Mishra, S.N. and Ramesh Chand (1995): "Public and Private Capital Formation in Indian Agriculture: Comments on Complementarity Hypothesis and Others", *Economic and Political Weekly*, vol. XXX, no. 25, June 24, pp. A-64-A-79.
11. Mitra, Ashok (1997): "Public and Private Investment in Agriculture" in Desai, Bhupat M. (eds.): *Agricultural Development Paradigm for the Ninth Plan under New Economic Environment*, CMA Monograph No. 179, Oxford and IBH Publishing Co. Pvt. Ltd., New Delhi, p. 976.
12. Dhawan, B.D. and S.S. Yadav (1995): *op. cit.,* p. A-109.
13. Dhawan, B.D. (1996a): "Relationship between Public and Private Investments in Indian Agriculture with Special Reference to Public Canals", *Indian Journal of Agricultural Economics*, vol. 51, nos. 1 & 2, Jan.-June, p. 211.
14. Alagh, Yoginder K (1997): "Agricultural Investment and Growth", *Indian Journal of Agricultural Economics*, vol. 52, no. 2, April-June, p. 282.
15. Gandhi, Vasant P. and Gyanendra Mani (1997): *Investment Behaviour in Indian Agriculture: A Theoretical and Empirical Investigation*, Working Paper, IIM Ahmedabad, p. 33.
16. Dhawan, B.D. (1996a): *op. cit.,* p. 215.
17. *ibid.,* p. 213.
18. The lists of 23 major heads of expenditure given in the finance account are given in the Chapter IV, Footnote 12.
19. Chand, Ramesh (2000): *op. cit.,* pp. 62-64.
20. The broad series defined by Gulati, Ashok and Seema Bathla (2000) included heads of capital expenditure given in budgetary documents and have already been discussed in Chapter IV, Footnote 14.
21. Gulati, Ashok and Seema Bathla (2000): *op. cit.,* p. 104.
22. Gandhi, Vasant, P. (1990): *op. cit.,* p. 61.
23. Gandhi, Vasant, P. (1996): *op. cit.,* p. 555.
24. Misra, V.N. and P.B.R. Hazell (1997): *op. cit.* p. 1992.
25. Misra, V.N. (1998): *op. cit.,* p. 2108.
26. Misra, V.N. and P.B.R. Hazell (1996): *op. cit.,* p. A-4.
27. Brown, Murry (1958): *On the Measurement of Technological Change*, Cambridge University Press, and Solow, Robert M. (eds.): "Technical Change and Aggregate Production Function",

Review of Economic and statistics, August, cited in Singh. W. Kumar (1987), *op. cit.*, p. 170.

28. Sidhu, D.S. and A.J. Singh (1986): "Technological Change in Indian Agriculture" in Dantwala, M.L. and Others (eds.): *op. cit.*, p. 140.
29. *ibid.*, p. 14l.
30. Misra. V.N. and Peter B.R. Hazell (1996): *op. cit.*, p. A-8.
31. Error term 'u_t' inter alia is a surrogate for all those variables that are not taken into account explicitly in the model but that collectively affect private capital formation. For more detailed analysis see Gujarati, Damodar N. (1995): *Basic Econometrics*, McGraw-Hill International, 3rd Editions, Singapore, pp. 39-41.
32. *ibid.*, p. 405.
33. *ibid.*, pp. 431-32.

6

Impact of Agricultural Capital Formation on Agricultural Production

6.1. Introduction

The importance of capital formation in realising the goal of raising production in every economic system has been well recognised. Agriculture sector is no exception. To reiterate further, the main objective of capital formation in agriculture sector is to provide impetus to growth of agricultural production. While public capital formation achieves this goal by creating, improving and maintaining suitable infrastructure such as rural roads, power, irrigation facilities, etc., private capital formation contributes by adding to stock of capital, durable assets, improvement of resources and efficiency of labour, etc. All these activities would augment productive capacity and also improve production efficiency of this sector.[1] Now the pertinent question is how much is the share of contribution of agricultural capital formation in the production of agricultural output?

To shed light on the role of capital formation in agricultural production and to ascertain its factor share, an empirical examination of agricultural production function with capital formation as one of the factor input is highly pertinent. Such analysis would enable us to know the contribution of different inputs to growth of agricultural output which has normally been judged by their contribution to output under constant technology, attributing the rest of the growth to technical change.[2] The assessment of agricultural production function would help the farmers to attain their stated objectives and the society to achieve the most efficient use of agricultural resources at their disposal. [3]

Prior to a brief literature survey on empirical studies regarding the issue under consideration, it would be highly pertinent to define what a production function is? "The production function is the mathematical counterpart of the applied term input-output relationship. Such a relationship may either be discrete or continuous."[4] In other words, a production function describes the technical relationship that transforms inputs (resources) into output (commodities).[5]

6.2. Review of Literature

There are a number of empirical studies on production function of agricultural output with capital formation as one of the factor input (Mallick, 1993; Chand, 2000; Gulati and Seema Bathla, 2000; Mundlak, Yair, et al., 2000 etc.). They attempt to ascertain the contribution of capital formation on agricultural output. Though the estimated input share of capital formation to agricultural output varies, their conclusion is unanimous that capital formation plays a major role in agricultural production. These variations in estimated input share may be mainly attributed to the difference in time period covered and the coverage and methodology adopted while constructing the time series of capital formation.[6]

Mallick (1993) attempted to estimate input share of

capital formation in agricultural output in Indian agriculture based on CSO series for the period 1960-61 to 1989-90. It would be worthwhile to reproduce two of his estimated linear regression equations.

1. NDPFCAGR = -18897 + 0.79655 GIA + 236.47 RAIN–$\bar{R}^2$ = 0.96
   (-3.6618) (10.274) (5.4068) DW = 1.94

2. GIA = 24335 + 2.7489 $GCFAGR_{-1}$ + 51.609 RAIN–$\bar{R}^2$ = 0.98
   (6.59) (4.3872) (1.5358) DW = 1.52

Where,

NDPFCAGR =	Net Domestic Product at Factor Cost originating from Agriculture at 1980-81 prices (Rs. Crore).
$GCFAGR_{-1}$ =	Gross Capital Formation in Agriculture lagged by one year (Rs. crore at 1980-81 prices)
GIA =	Gross Irrigated Area (in thousand hectares)
RAIN =	Rainfall index (100 = normal)

The first equation revealed that the NDPFCAGR is positively influenced by GIA and RAIN at 1% level of significance. While the second equation revealed that GIA is positively influenced by one year lag GCFAGR and RAIN. But only one-year lag GCFAGR is significant at 1% level of significance. Thus, based on these two estimated equations he concludes that agricultural output is significantly influenced by investment in irrigation which constitutes about 90% of public capital formation.[7]

When a number of scholars questioned the definitional aspect of CSO estimates of agricultural capital formation particularly public capital formation, its role on growth of agricultural output has also started to looked afresh. Ramesh Chand (2000) was the first to work in this direction. He tried to examine the nature of relationship between agricultural productivity and capital formation across the states of Indian union along with other relevant variables at 1980-81 prices with the help of multiple regression analysis. His analysis is based on the broad series[8] of public investment constructed by him. Further, he attempted to ascertain the relative importance of the role played by public and private investment in the growth of agricultural output. The analysis revealed that private investment exerted stronger impact than public investment on inter-state variation in agricultural productivity. This is shown by the positive but insignificant effect of public investment. But when the two series (public and private investment) are combined as one variable, it showed significant positive impact on inter-state variation in output. These results confirmed that the level of agricultural productivity and output growth rate are significantly affected by the level of fixed investment in different states. Thus, he concluded that agricultural investments are an important factor input to the growth of agricultural productivity and hence income.

He also attempted to examine the inequalities in inter-state variation in agricultural investments by comparing the coefficient of variation over different plan periods. The estimated coefficient of variation in public capital investment among major states of India declined from 68% during 5th plan period to 48% during 7th plan period and then to 47% during the following two years (1990-91 to 1991-92). It, however, jumped to 70% during the 8th plan period. A similar picture could also be seen in the case of private investment when it declined from 65% in 1981-82 to 55% in 1991-92.[9] Further, the incremental capital output ratio (ICOR) for agricultural sector for last 20 years was also calculated based on both CSO and broad series. The estimated ICOR based on both broad series and CSO series shows a declining trend till 1989-90 then it increases in the subsequent five years even though there is a huge difference in the estimated ICOR between the two series in the beginning but slowly narrowed down overtime. Accordingly the marginal efficiency of capital in agriculture improved till 1989-90 and deteriorated in the subsequent quinquennium. He commented that the planning commission's ICOR is an underestimate when compared with the results of his study.[10]

Similar attempt has also been made by Gulati and Seema Bathla (2000) based on both conventional CSO series and broad series[11] in terms of both physical and financial form[12] at 1993-94 prices for the period 1980-81 to 1998-99. While doing so, they have also taken into account of the impact of terms of trade and gross cropped area. To examine the impact of these factor inputs on growth in agriculture they resorted to reduced form method/indirect least square method. They opined that the private capital formation in agriculture, that influence gross domestic product in agriculture (GDPA) depends on public capital formation and terms of trade, has been considered as an endogenous variable. Therefore, the impact of terms of trade and public capital formation on GDPA cannot be measured directly by any of the structural parameters but only through simultaneous solution of the system.

Their estimated equation revealed that with 10% increase in cumulative public gross capital formation in irrigation work, GDPA would increase by 23% when irrigation is considered in cumulative physical form i.e. as canal intensity and by 2.53% when considered in cumulative financial terms. But when the public investment in power supply to agriculture sector is added to canal irrigation the percentage increase in GDPA is reduced to 1.6% and 2.43% when considered in physical and financial form respectively. It is, however, found that there is not much variation in the elasticity of GDPA with respect to broad series of public investment.[13]

Though there is slight variation in the estimated elasticity of GDPA with respect to public capital formation in agriculture according to its coverage and forms (physical and financial) under consideration, "public sector capital formation in agriculture, which ever form it is considered is an important variable in influencing growth in agricultural GDP. It affects agricultural GDP directly, as well as indirectly through its inducement effect on private sector investment in agriculture. Since in a structural form of equations private sector investment is strongly influènced by public sector investments and terms of trade, it becomes an endogenous variable. Thus, in a reduced form, impact of private sector investments on agricultural GDP is subsumed by the impact of public sector investment and terms of trade on agricultural GDP".[14]

A cross section analysis on the role of capital formation in agricultural production has been conducted by Mundlak et al. (2000). They attempted to estimate a cross-country agricultural production function with land, labour, capital and fertilizers and pesticides as factor inputs with the help of multiple regression analysis. While doing so they took into account of annual data from 37 countries for the period of 21 years (1970-90). The analyses revealed that capital play a significant role in the growth of agricultural output. This was shown by the estimated coefficient of capital that accounts for about 40% of total output. Thus, they concluded that agricultural technology is cost-capital intensive as compared to non-agricultural sector. According to them a technology is cost-capital intensive with respect to a reference technology if the factor share of capital is larger than that of the reference technology.[15]

The empirical analysis discussed above reveals that capital formation plays an important role in the growth of agricultural output. Any short fall in agricultural capital formation would invariably lead to decline in agricultural output. Hence proper assessment of the level of capital formation and its role in achieving sustainable agricultural development[16] is highly pertinent.

So far no attempt has been made to examine and analyse this important issue for the state of Manipur. Singh (1985), however, attempted to examine the production function of rice crop for the period 1951 to 1985 without taking into account the role of capital formation. He considered rainfall during the plantation period (R_{1t}) and growing period (R_{2t}), the consumption of fertilizers (F_t), area under HYV (H_t), the area under the rice crop (A_t) and the trend variable (T) as factor inputs as they are assumed to affect the production of rice crop in Manipur. He used the multiple regression analysis of different models viz., linear, log-linear and transcendental equations. The estimated regression equations revealed that the most important explanatory variable in explaining the variation in rice output is rainfall during the plantation period which is followed by rainfall during growing period, area under the rice crop, and HYVs and consumption of chemical fertilizers respectively.[17]

As stated earlier, any empirical work on the role of capital formation in agricultural production for the state of Manipur is yet to be done. The present study, therefore, is an attempt to examine this critical issue by fitting production function of agricultural output in Manipur during the period 1972-73 to 1997-98.

6.3. Specification of Variables

It is hypothesized that the production of agricultural

output in Manipur is a function of land, labour, capital formation and the level of technological development in the state.[18] The extent or level of technological development in the state is represented by the extent of HYVs diffusion and consumption of chemical fertilizers. The implicit functional form that can be used in fitting agricultural production function is;

$$Q_t = f(L_{1t}, L_{2t}, K_t, H_t, F_t)$$

Where, Q_t = NSDP originating from agricultural sector at 1980-81 prices (Rs. in Lakh).

L_{1t} = Total area under the cultivation of rice crop in the state (in thousand hectares).

L_{2t} = Total number of persons engaged as agricultural worker/labour (in thousand)

K_t = Capital formation in agricultural sector at 1980-81 prices (Rs. in lakh)

H_t = Area under HYVs of rice crop (in thousand hectares)

F_t = Amount of chemical fertilizers consumption in materials (Urea, DAP and MOP in thousand tons)

The detailed specifications of the different explanatory variables included are as follows:

6.3.1. Land (L_{1t})

The hectares of area under rice crop have been taken into account as a measure of land while explaining the variation in agricultural output in the state. The inclusion of area under rice crop as a factor input is expected to capture two contrasting aspects. First, the negative growth rate of area under rice crop (-0.48) and resulting negative area effect from decomposition of growth of rice output during 1972 to 1997 may lead to decline in output of rice crop in the state.[19] Secondly, the development of irrigation facilities and other infrastructure and modern technology may not only restore to the land its inherent powers but may also contribute in a great way to improve these powers to raise higher levels of output from a given stock of land. It may have offset the effect of negative growth rate of area under rice crop thereby resulting into positive growth rate of rice output. Thus, such analysis would enable us to know whether the present state of technology is land saving or labour saving.

6.3.2. Labour (L_{2t})

Labour, whether in traditional or modern agriculture, constitutes an important factor of production. Machine tools and labour cannot be perfect substitutes in the production process. The growth and development of this sector is highly dependent on the productive efficiency of labour. In other words, the development of agricultural sector is directly related with the size and quality of labour force and the nature and level of technological development, which are directly related to each other.

The discovery of new machines and techniques by the agricultural scientists alone would not increase the production if they were not backed by innovative skills and efficient farm labour. The use of new technology warrants new knowledge among the users. In any reform, technology may remain idle if the knowledge to use it has not been developed simultaneously and diffused among farmers. While efficiency of farm labour can be measured in a simple way, by the amount of output produced per unit of labour, quality of labour forces measured by physical health, educational qualification, technical training, etc. lacks adequate data.

The present study, therefore, shall examine only the effect of the size of labour forces and the level of technological development on growth of rice output. It has been widely viewed that in traditional agriculture increase in labour force tends to reduce its output while the opposite is true in the case of modern agriculture. It means the effect of the size of labour force is determined by the nature and level of technological development. Hence the size of labour force has been included in the production function to capture its effect on the rice output.

Agricultural labour is defined as the economically active population in agriculture.[20] The data on the size of farm labour is obtained from various census reports. Thus, the census report of 1970-71, 1980-81 and 1990-91 were used as benchmark for the estimates of other years. In other to obtain the time series on labour force, given in Appendix 6.1A, we took the help of compound growth rate of these benchmark years to interpolate and extrapolate for the intervening years and to extend beyond 1990-1991.

6.3.3. Capital Formation (K_t)

Agricultural capital formation means net addition made to the existing stocks of capita-produced means of further production-in a given period of time. As it plays an important role in improving agricultural production, increase in capital formation in agricultural sector would exert positive and significant impact on production of agricultural output.

6.3.4. Technology

The role of technology in the process of agricultural production has already been discussed briefly in previous Chapter V. To reiterate further, whatever the changes that have been taking place in the forms and nature of capital conducive to the existing institutional framework and environment of that particular region/state and development of new techniques have been the consequences of technological progress.

The development of new technology raises the level of output that can be produced from a given amount of factor inputs. In other words, the increase in the productivity of factor inputs is mainly attributed to technology component of production function. Thus, development of new technology would significantly influence the output of rice in the state. The extent of technological diffusion or the spread of technological adoption included in the equation is, therefore, expected to capture the contribution of technology in the production process. From the technological basket[21] only the area under HYVs and consumption of chemical fertilizers are, for the present study, used as a proxy for technology. This is mainly due to the lack of adequate information on other components included in the basket.

6.3.4.a. Area under High Yielding Varieties (H_t)

The area under high yielding varieties has been introduced in the function, representing a component of technology, as HYVs technology is known to have significant impact in the output of rice crop. The area under HYVs of rice crop has increased from a mere 7 thousand hectares in 1972-73 to 92 thousand hectares in 1991-92. Since then it slide down somewhat and hovered around 70 to 72 thousand hectares. The time series on area under HYV paddy is given in Appendix 6.1B. It means the diffusion of HYVs declined in the 90's instead of increasing. This may have some adverse impact on rice output in the state.

6.3.4.b. Fertilizer Consumption (F_t)

The consumption of chemical fertilizers in the state has been increasing continuously since 1972-73. It increased from 1.31 thousand tonnes in 1972-73 to 13.19 thousand tonnes in 1997-98. The time series on fertilizer consumption in Manipur is given in Appendix 6.1B. This increase in the level of fertilizer consumption has been attributed to two reasons (Singh, 1985). First, extension of area using chemical fertilizers-which we have witnessed in the seventies. Second, increase in the intensity of using chemical fertilizers as in the eighties and nineties.[22] The past experience of Indian agriculture revealed that increasing application of chemical fertilizers coupled with widespread HYVs adoption has contributed substantially to the growth of agricultural output. Hence the increase in fertilizer consumption in the state is, therefore, expected to exert significant impact on the output of rice.

The increase in intensity of fertiliser consumption would also have some negative impact in the long run. Such increase would lead to environmental degradation and slow deterioration in soil health. Similarly, the residue of these chemicals in food product would also affect both human and livestock health. These would adversely affect the productivity of soil, human and livestocks. Thus, the inclusion of fertilizer consumption in the production function is expected to capture these contrasting effects.

6.3.5. *Error Terms (u_t)*

Error term 'u_t' represents the combined influence of all those variable that are not included explicitly as independent explanatory variable in the model but which collectively affect the output of rice in the state. The exclusion of these variables become plausible due to a number of reasons viz. unavailability of data, poor proxy variables, principle of parsimony, wrong functional form and joint influence of some variables and peripheral variables may be very small.[23]

6.4 Model Specification

For empirical estimation of production of agricultural output in the state, the non-linear Cobb-Douglas production function has been considered for the present study. We choose this particular model as a number of macro-modellers viewed this specification as more useful in empirical estimation of production function of agriculture rather than industry.

The Cobb-Douglas production function, in its stochastic form, may be expressed as :

$$Q_t = a_0 L_{1t}^{b1} L_{2t}^{b2} K_t^{b3} H_t^{b4} F_t^{b5} e^{ut}$$

Since Cobb-Douglas production function is non-linear in parameter we are, therefore, required to transform the model into linear form in order to apply regression method. By taking natural log on both the side of the above equation/ model we would get the desire form of linearity. Hence we obtain,

$$\text{In } Q_t = \text{In } a_0 + b_1 \text{In } L_{1t} + b_2 \text{In } L_{2t} + b_3 \text{In } K_t + b_4 \text{In } H_t + b_5 \text{In } F_t + u_t$$

Or,

$$\text{In } Q_t = b_0 + b_1 \text{ In } L_{1t} + b_2 \text{ In } L_{2t} + b_3 \text{In } K_t + b_4 \text{In } H_t + b_5 \text{In } F_t + u_t$$

Where, $b_0 = \text{In } a_0$

The model is linear in parameters with b_0 as the intercept term and b_1, b_2, b_3, b_4 and b_5 are the coefficients of the variables and u_t represents the error term. Such transformation would enable us to apply ordinary least square method of regression.

6.5 Findings

The degree of association between agricultural output represented by Net State Domestic Product (NSDP) originating from agricultural sector at 1980-81 prices in the state and other relevant explanatory variables considered in the present study during the period 1972-73 to 1997-98 is shown by the estimated coefficients of correlation which is presented Table 6.1.

TABLE 6.1

Correlation between Output and Other Relevant Variables

	Q_t	L_{1t}	L_{2t}	K_t	H_t	F_t
Q_t	1.00	- 0.1467	0.8672	0.7241	0.6640	0.8326
L_{1t}		1.00	- 0.4143	- 0.1674	- 0.1992	- 0.4674
L_{2t}			1.00	0.7181	0.8142	0.9468
K_t				1.00	0.7879	0.6407
H_t					1.00	0.6380
F_t						1.00

Note: Absolute value of estimated correlation coefficient above 0.5368, 0.4227 and 0.3599 are significant at 1%, 5% and 10% level of significance respectively.

The above Table shows that the estimated coefficient of correlation 'r' between agricultural output (NSDP) and land is found to be negative but insignificant (-0.1467). The negative value of r may be mainly due to downward revision of widely fluctuating trend of area under rice cultivation (land) while NSDP originating from agriculture is ever growing. However, the negative correlation needs to be elucidated further. While for other remaining combinations viz. output and labour, output and GCF, output and HYVs, and output and fertiliser consumption the estimated correlation coefficients are 0.86717, 0.7241, 0.66404 and 0.83256 respectively. They are all positively significant at 1% level of significance. This means that the agricultural output in the state is positively associated with these four variables.

High correlation coefficient implies that the agricultural output in the state may be significantly affected by these variables. In order to examine the effect of these variable on agricultural production both log-linear simple and multiple regression analysis were carried out. Let us first examine the results of the log-linear simple regression as shown in Table 6.2.

TABLE 6.2

Simple Log-Linear Regression between Output and Other Relevant Variables

Sl. No.	*Independent Variable*	*Estimated Regression Coefficient*	$\overline{R}^2$	*DW*
1.	Log (L_{1t})	0.49185 (2.257)**	- 0.069	1.7379
2.	Log (L_{2t})	0.36435 (2.391)**	0.6938	1.9673
3.	Log (K_t)	0.34609 (4.909)*	0.667	1.9781
4.	Log (H_t)	0.24933 (5.413)*	0.6387	2.2648
5.	Log (F_t)	0.22941 (4.761)*	0.6713	1.6844
6.	Log ($PUBGCF_t$)	0.27632 (4.353)*	0.618	2.0376
7.	Log ($PVTGCF_t$)	0.63974 (7.045)*	0.7787	1.8413
8.	0.14968 Log ($PUBGCF_t$) + 0.44275 Log ($PVTGCF_t$) (2.716)** (4.469)*		$\overline{R}^2$ = 0.826	DW =1.7363

Note: 1. * & ** Significant at 1% and 5% level of significance respectively.
2. Figures in the parenthesis indicates t ratio.

When each of the five variables is used independently as an explanatory variable, all the specifications suffer from the problem of autocorrelation. The presence of autocorrelation problem may be attributed to sluggishness or inertia[24] in the time series data of the variables. The presence of autocorrelation affects the test of significance of the estimated coefficients of the ordinary least square

(OLS). It, therefore, is highly pertinent to correct the problem with the help of some standard remedial measures. Here the problem has been corrected with the help of Prais-Winsten Technique.

The estimated simple log-linear regression coefficients of all the five variables are positive and significant at either 1% or 5% level of significance. The equation no. 1, shows that even though the elasticity of land is high (0.492) its explanatory power is very low (-0.0369). High magnitude of elasticity implies that land continues to be one of the most important inputs in agricultural production. While the magnitude of elasticity of four remaining variables (equation 2 to 5) varies between 0.364 and 0.229 and their explanatory power varies between 69% and 64%. The result of the simple log-linear regressions implies that land, labour, GCF and technological development positively affects the growth of agricultural output in the state.

Further, an attempt has also been made to determine the relative importance of capital formation originating from two sector, viz. public and private sector, on agricultural output. The results of the regression of output on public and private capital formation independently are shown in equation nos. 6 and 7. The equation no. 6 revealed that the elasticity of public capital formation is 0.276 and explained nearly 62% of the output variation. But when output is regressed on private capital formation, it explained nearly 78% of the output variation. While the magnitude of elasticity of private capital formation is 0.64. Thus, the results indicate that private capital formation is relatively more important in explaining the variation in agricultural output. In other words, private capital formation plays more prominent role in the growth of agricultural output.

Again, when the same is regressed on public and private

capital formation simultaneously nearly 83% of the variation in agricultural output has been explained. The elasticity of agricultural output with respect to pubic capital formation is 0.15 and with respect to private capital formation is 0.44. It implies that 10% increase in public capital formation would lead to a 1.5% increase in agricultural output. While the same proportion of increase in private capital formation, agricultural output would increase by about 4.4%. The above results (given in equation 6 to 8) indicate that the private capital formation plays relatively more important role in the growth of agricultural output than public capital formation.

Like other sectors of the economy production of agricultural goods is possible when various inputs/variables are put together simultaneously. We examine this by using the Cobb-Douglas production function using all the five explanatory variables viz. land, labour, GCF, HYVs and fertilizer consumption. The results show the presence of autocorrelation problem. After the problem has been corrected with the help of Prais-Winsten Technique, the result revealed that all the estimated coefficients bear the expected positive sign. But, other than land which is significant at 10% level of significance, all the other remaining variables are insignificant. However, the estimated value of F (=19.421) is significant at 1% level of significance. It is observed that existence of high correlation between dependent and different independent variables and F-test being significant at 1% level of significance indicates the possibility of the result being due to multicollinearity. With F-test being significant at 1% level, insignificance of estimated coefficients of variables may be mainly due to the presence of high degree of multicollinearity.

The presence of multicollinearity problem could be

confirmed when we look at the estimated coefficient of correlation among the explanatory variables given in Table 6.1. The Table shows that all the estimated coefficients of correlation between land and other explanatory variables i.e. Labour, GFC, HYVs and fertilizers consumption are negative. However only two of them i.e. land and labour and land and fertilizer consumption are significant at 10% and 5% level of significance respectively. While the estimated coefficient of correlation between other remaining variables are positive and significant at 1% level. The result indicates that there exists high correlation among the different explanatory variables. High degree of correlation among different explanatory variables implies the overlapping of the variables thereby leading to the problem of multicollinearity.

Nevertheless, an attempt has been made to examine the combined effect of these variables on agricultural output dropping alternate variables and choose the best-fit model. The results of the multiple regression analysis with their summary statistics, $\overline{R}^2$ and corrected value of autocorrelation using the range of data from 1972-73 to 1997-98 are presented in Table 6.3.

Equation I of the Table 6.3 show that when land, labour and GCF are used as explanatory variables nearly 80% of the output variation has been explained. The estimated coefficients of land and GCF are significant at 5% level and for labour at 1% level of significance. When we look at the magnitude of elasticity, GCF is the third major input after labour and land in the production of agriçultural output. That is, 10% increase in GCF would lead to 1.53% increase in output as against 6.166% and 4.591% for labour and land respectively. Equation II and III revealed that, when GCF is dropped from the equation, the explanatory power is reduced to 76% and 77% respectively. In the former labour and in the latter land occupy the most important factor input required. The contribution of HYVs slightly improved when fertilizer consumption is included in the equation III. Again when GCF is considered as regressor along with land and fertilizer consumption their explanatory power increased to little over 80% of the output variation (equation IV). The elasticity of GCF has also increased to 0.16879. Similar to equation I, GCF continued to be the third important factor input required in agricultural production.

TABLE 6.3

Log-Linear Multiple Regression between Output and Different Combination of Various Explanatory Variables

Sl. No.	*Log* (L_{1t})	*Log* (L_{2t})	*Log* (K_t)	*Log* (H_t)	*Log* (F_t)	R^2	*DW*
I.	0.45909 (2.278)**	0.61662 (3.678)*	0.1527 (2.009)**			0.7973	1.931
II.	0.36832 (1.911)***	0.53228 (2.154)**		0.13209 (2.079)**		0.7578	2.1403
III.	0.42805 (2.059)**			0.14058 (2.431)**	0.14247 (2.26)**	0.7710	2.1389
IV.	0.53431 (2.499)**		0.16879 (2.306)**		0.17345 (3.709)*	0.802	1.9624
V.	0.53537 (2.71)**	0.88732 (5.897)*				0.7598	1.6756
VI.		0.50183 (2.98)*	0.20273 (2.586)**			0.7737	1.6845
VII.			0.23069 (3.065)*		0.12829 (2.77)**	0.7705	1.7459
VIII.	0.67601 (3.17)*				0.2511 (6.197)*	0.7588	1.8951

Note: 1. *, ** and *** Significant at 1%, 5% and 10% level of significance respectively.

2. Figures in the parenthesis indicates t ratio.

Further, when only land and labour (equation V) or land and fertilizer consumption (equation VIII) are considered, their explanatory power reduced to around 76 percent. Similar to equation I and II, labour accounts for the largest share in the former (equation V) and land in the latter (equation VIII). However, when labour and GCF in equation VI and GCF and fertiliser consumption in equation VII are considered, explanatory power further improved to little over 77%. In these two equations the elasticity of GCF also increased to 0.20 in the former and 0.23 in the latter.

All these equations revealed that labour, land and GCF play dominant role in the growth of agricultural production while technology plays subordinate role. If we are to pick the best-fit model, equation IV is the best-fit when land, GCF and fertiliser consumption are fitted simultaneously and equation I is the second best. In these best-fit models the elasticity of GCF is 0.16879, which implies that with 10% increase in the capital formation, agricultural output would be increased by about 1.69 percent. In other words land, labour and capital formation are the most indispensable primary inputs of the agricultural production.

Further, in the same line of our earlier argument private capital formation plays more dominant role in the growth of agricultural production, we shall carry forward the same in multiple regression analysis frameworks as well. Here instead of using GCF we shall use either public or private capital formation series only while other variables remain the same. The results of the multiple regression analysis with their summary statistics, $\overline{R}^2$ and corrected value of autocorrelation are presented in Table 6.4.

TABLE 6.4

Log-Linear Multiple Regression between Output and Segregated series of Capital Formation (Public & Private) with Other Relevant Variables

Sl. No.	*Log(PUBGCF$_t$)*	*Log(PVTGCF$_t$)*	*Log-(L_{1t})*	*Log-(L_{2t})*	*Log-(H_t)*	*Log-(F_t)*	R^2	*DW*
I.	0.12855		0.43909	0.64097			0.7934	1.9458
	(2.063)**		(2.196)**	(3.931)*				
II.	0.1414		0.51717			0.18059	0.7996	1.9693
	(2.353)*		(2.43)*			(4.015)*		
III.	0.17015			0.53828			0.7721	1.7214
	(2.676)**			(3.283)*				
IV.	0.19269					0.13906	0.7702	1.7733
	(3.155)*					(3.138)*		
V.		0.56002	0.6126		0.062304		0.8744	1.7164
		(6.831)*	(2.961)*		(2.048)**			
VI.		0.68891	0.77009				0.8569	1.738
		(12.284)*	(3.757)*					

Note: 1. * & ** Significant at 1% and 5% level of significance respectively.
2. Figures in the parenthesis indicates t statistics

Equation 1 of Table 6.4 revealed that when public capital formation, land and labour were used as regressors, 80% of the variation in agricultural output was explained. The estimated coefficients of all the three variables are positive and significant at 10% level (Public capital formation), 5% level (land) and at 1% level (labour) of significance. The elasticity of public capital formation is 0.12855 while that of land and labour are 0.43909 and 0.64097 respectively. This implies that 10% increase in public capital formation would lead to 1.29% increase in the agricultural output. In the case of land and labour, agricultural output would increase by 4.39% and 6.41% respectively. In equation II, the explanatory power of the model and the elasticity of public capital formation improved when agricultural output is regressed on public capital formation, land and fertilizer consumption. Similarly, equation III and IV revealed that when public capital formation was used as regressor either with labour or fertilizer consumption only, its elasticity further improved to 0.17015 and 0.19269 respectively. However, the explanatory power of the two models (equation III & IV) declined to 77 percent.

On the other hand when private capital formation, land and HYVs are used as explanatory variables, the explanatory power of the model rose to 87 percent of the output variation (equation V). The estimated coefficients of private capital formation and land are significant at 1% level and at 5% level in the case of HYVs. The elasticity of land and private capital formation are 0.6126 and 0.56002 respectively while the elasticity of HYVs is the lowest (0.062304). It means 10% increase in private capital formation would lead to 5.6% increase in agricultural output. While same proportion is increase on land and HYVs, agricultural output would increase by about 6.13% and 0.623% respectively. In equation VI, the explanatory power of the model is slightly reduced to 86% when private capital formation and land were used as explanatory variables.

But their elasticity improved to 0.68891 for former and 0.77 in the case of latter. It means 10% increase in private capital formation or land would lead to 6.9% or 7.7% increase in agricultural output.

The above results implied that 10% increase in public capital formation would bring about less then 2% increase in output. While the same amount of increase in private capital formation would lead to 5 to 6% increase in output. The obvious reason of low rate of return from public capital formation may be mainly due to the long gestation period before the realization of actual benefits. At the same time the very nature of public goods, return to public capital formation may also be small which linger throughout the life span of the capital. On the other hand farmers are more interested in investing where the return is large and immediate. Thus, we can conclude that capital formation-whether originating from public or private sector- constitutes an important factor input required in the growth of agricultural output.

Further, an attempt has also been made to determine the magnitude of returns to scale. By returns to scale we mean the rate at which output increase as the quantities of all factor inputs are increased simultaneously in the same proportion. It is generally measured by the ratio of percentage increase in output divided by percentage increase in input, under the condition that all factors are increased by the same proportion.[25] It means returns to scale is equal to the sum of elasticity of production with respect to the different inputs. Thus, in the case of Cobb-Douglas production function, as used in our present analysis, returns to scale is the sum of all the bi's which are the coefficients of different explanatory variables. If the sum of bi's is equal to 1 i.e. $\Sigma bi = 1$, then there is constant returns to scale. It means doubling the inputs will double the output, tripling the inputs will triple the output and so on. If the sum is less than 1 i.e. $\Sigma bi < 1$, there is decreasing returns to scale - doubling the inputs will raise output less than double. If the sum is greater than 1 i.e. $\Sigma bi > 1$, there is increasing returns to scale. Doubling the inputs will raise output more than double. The procedure normally adopted by macro-modellers to ascertain if the sum of bi's is significantly different from unity at the desired probability level is to perform a t-test.[26] For our present analysis the magnitude of returns to scale has been tested only for two best fit model chosen from different set of equation given in Table 6.3 and 6.4 i.e. equation IV from Table 6.3 and equation V from Table 6.4.

It has been hypothesized that agricultural production in the state is characterised by constant returns to scale. An equiproportional change in output for an equiproportional change in the inputs i.e. $b_1 + b_2 + b_3 = 1$. The t-test[27] has to be performed. If the computed t value exceeds the critical t-value (table value of t) at the chosen level of significance, we may reject the hypothesis of constant returns to scale; otherwise we may not reject it. In the first case, when output is regressed on land, GCF and fertilizer consumption (equation IV Table 6.3) the computed t-value (t=0.5792) is less than the critical value even at 10% level. Hence, we accept the hypothesis that there is a constant returns to scale implying thereby an equiproportional increase in output when land, GCF and fertiliser consumption are increased by same proportion.

In the second model, i.e. when output is regressed on land, private capital formation and HYVs (equation V, Table 6.4), the computed value (t=3.2713) is more than the table value (2.807) even at 1% level of significance. Hence we reject the hypothesis of constant returns to scale. In other words, there is either decreasing or increasing returns to scale when land, private capital formation and HYVs are used as explanatory variable. This implies that whenever, there is increased in land, private capital formation and HYVs in same proportion, output will be increase either less then or more than the proportionate increase in factor inputs. Thus, from these two models we can conclude that if we are to increase our agricultural output more than some proportionate increase in inputs we must consolidate private capital formation and invest more on technological development via increase in public capital formation. In short, to break the constant return to scale nature of production we must increase the rate of capital formation.

6.6. Analysis of Field Survey Data

A cross-section analysis on agricultural production has also been conducted at micro-level study based on field survey data to substantiate our earlier macro-level findings. While doing so we attempt to fit the Cobb-Douglas production function. However, out of the five variables employed as explanatory variable at macro level study only four of them could be extracted from the survey data. The fifth variable–area under HYV of rice crop–could not be extracted as the entire households covered in the field survey have adopted HYVs of rice crop. At the same time total rice output measured in kilogram were used as dependent variable instead of NSDP originating from agricultural sector. Further, labour and chemical fertilizer or organic manure consumption were in financial variable. Otherwise the relationship between dependent variable and other explanatory variables remained same with macro level study. The details of the variables under consideration are as follows:

6.6.1. Land (L_1)

The local measurement of households agricultural land holdings is called *'Sangam'*. But for the present study it has been converted into acre. The official equivalent of one *'Sangam'* is 0.62 acre.

6.6.2. Labour (L_2)

The total wage cost on labour force is used instead of the number of mandays labour mainly because wages of a hired labour is generally paid according to area i.e. per *'Sangam'* irrespective of number of hours he or she works. The total labour force employed in rice cultivation comprises of family labour and hired labour. So to arrive at the total wage cost, wage rate paid to the hired labour in a particular locality/ household has been used as imputed rate.

6.6.3. Capital formation (K_1)

As stated earlier household capital formation could not be extracted from the field survey data. We, therefore, use the stocks of capital accumulated by households as a proxy variable for private capital formation. The details of the quantum, structure and composition of stocks of capital accumulated at farmers account have already been explained in Chapter-IV. We, therefore, need not reiterated further.

6.6.4. Fertiliser Consumption (F_t)

There are a number of households who used organic manure instead of chemical fertilisers. Their prices are generally fixed in per bullock cart. Therefore, in order to formalise the scale of measurement we considered the prices at which both chemical fertiliser and organic manure are brought to arrive at the total cost of procuring chemical fertiliser and organic manure.

6.7. Findings

The degree of association between agricultural output and explanatory variables can be seen from the estimated coefficients of correlation which are given in Table 6.5.

TABLE 6.5

Correlation Coefficient between Different Explanatory Variables

Variable	*Output*	L_1	L_2	K_1	F_t
Output	1.000	0.964*	0.90944*	0.54097*	0.86272*
L_1		1.000	0.93354*	0.51742*	0.85743*
L_2			1.000	0.52766*	0.78711*
K_1				1.000	0.53065*
F_t					1.000

Note: * Significant at 1% level.

Table shows that all the estimated coefficients of correlation (r) between output and each of the explanatory variables bear positive sign and are significant at 1% level of significance. The high correlation coefficients imply that agricultural production is closely associated with these four explanatory variables. In order to examine the effect of these variables on agricultural production both log-linear simple and multiple regression analysis were carried out. The summary results of the log-linear simple regression are given in Table 6.6.

TABLE 6.6

Log linear Simple Regression

Sl. No.	*Independent Variable*	*Estimated Regression Coefficient*	R^2	*DW*
1.	Log (L_1)	0.93165 (57.498)*	0.9154	2.0608**
2.	Log (L_2)	0.86784 (37.042)*	0.7808	2.2326**
3.	Log (K_1)	0.2298 (10.396)*	0.2676	1.9586
4.	Log (F_1)	- 0.00024834 (-0.418)	- 0.2825	1.8336

Note: 1. * Significant at 1% percent level of significance.
2. ** Autocorrelation problem has been corrected with the help of Cochrane-Orcutt Iterative process.
3. Figure in the parenthesis indicates t statistics.

When each of these four explanatory variables is used independently as an explanatory variable, other than fertiliser and organic manure consumption, the estimated coefficients of three other remaining variables bear positive sign and are significant at 1% level of significance. The positive and significant coefficients imply that any variation on these variables will have direct bearing on agricultural output. The estimated coefficient of fertiliser consumption turns out to be negative though not significant. The negative coefficient implies that further increase in fertiliser application would lead to decline in agricultural output. However, this negative relation between agricultural output and fertiliser consumption needs further scrutiny.

When all the four explanatory variables are used simultaneously as a regressor, only land and accumulated capital stock on farmers account are significant at 1% level of significance while the other two variables viz. labour and fertiliser consumption are insignificant. With F-test being significant at 1% level the observed result being due to the presence of multicollinearity problem.

The presence of multicollinearity problem could be confirmed when we look at the estimated coefficients of correlation among the explanatory variables which is given in Table. 6.5. The Table shows that all the estimated coefficient of correlation among the four explanatory variables bear positive sign and are significant at 1% level of significance. The high degree of correlation among different explanatory variables implies the overlapping of the variables thereby leading to the problem of multicollinearlity.

Nevertheless, an attempt has been made to examine the combined effect of these variables on agricultural production dropping alternate variable and choosing the best-fit models from aggregate and district level which are given in Table 6.7.

For aggregate, the chosen specification indicates that land and accumulated capital stock of farmers explain nearly 92% of the output variation. The two estimated coefficients are positive and significant at 1% level of significance.

TABLE 6.7

Log Linear Multiple Regression for Aggregate and District Level

	Log (L_1)	*Log* (L_2)	*Log* (K_1)	*Log* (F_1)	R^2	*DW.*
All	0.9021 (49.805)*		0.028721 (3.451)*		0.9188	2.0534+
ImphalEast	0.78335 (16.449)*			0.16783 (3.338)*	0.9272	2.0142
ImphalWest	0.86182 (21.874)*		0.053188 (3.47)*		0.9171	1.9866
Bishnupur	0.66931 (8.489)*	0.15942 (1.964)***		0.12194 (2.552)***	0.9324	1.9009
Thoubal	0.89864 (24.415)*		0.033505 (1.766)***		0.9356	2.0782

Note: 1. * , ** & *** significant at 1%, 5% and 10% level of significance respectively.

2. Autocorrelation problem has been corrected with the help Cochrane-Orcutt iterative process.

3. Figures in the parenthesis indicates t statistics.

This is to say that 10% increase in fixed investment on formers' account would lead to 0.29% increase in agricultural output. The results of the multiple regression analysis imply that land and capital stock at farmers account play a significant positive role on variation in output. This, however, does not mean that other variables do not have any significant role to play in increasing the production of agricultural output. This could be seen when we examine the best-fit models extracted from the analysis at district level as shown in Table 6.7.

In Imphal East district the specification with land and fertiliser consumption provides the best fit. The two variables explain nearly 93% of the output variation and their estimated coefficients are significant at 1% level of significance. The estimates of elasticity of output with respect to land and fertiliser consumption are 0.78335 and 0.16783 respectively. In Imphal West district the specification with land and capital stock is the best fit and the two variables explain nearly 92% of the output variation. The coefficients of the two variables bear expected positive sign and are significant at 1% level of significance. The estimates of elasticity of output with respect to land and capital stock are 0.86182 and 0.053188 respectively.

In Bishnupur district the specification with land, labour and fertiliser consumption provides the best fit and the three variables explain more than 93% of the output variation. However, the level of significance for the estimated coefficient of labour is very low (significant at 10% level of significance). The estimates of elasticity with respect to land, labour and fertiliser consumption are 0.66931, 0.15942 and 0.12194 respectively. The result of Thoubal district is similar to that of Imphal West district but their explanatory power is slightly improved to about 94%. However, the level of significance for estimated coefficient of capital stock is low (significant at 10% level of significance). From the above results we can conclude that all the four explanatory variables have significant role to play in the growth of agricultural production. The negligence on any one of these variables would lead to decline in agricultural production.

Further, an attempt has also been made to determine the magnitude of the returns to scale of these selected specifications with the hypothesis that agricultural production in the valley areas of the state follows constant returns to scale i.e. $\Sigma bi = 1$. For this we performed F-test.[28] If the computed F value exceeds the critical F value (table value of F) at the chosen level of significance, we may reject the hypothesis of constant returns to scale; otherwise we may not reject it.

The result of the aggregate data (equation 1, Table 6.7), the computed F value (F=1427.765) exceeds the critical value of F (=6.63) at 1% level of significance and hence rejects the hypothesis of constant returns to scale. In other words, in the valley area as a whole agricultural production follows either increasing or decreasing returns to scale. This result corroborates our earlier finding at macro level study where land, private capital formation and HYVs were used as explanatory variables.

In the district level, three districts viz Imphal East Imphal West and Thoubal district exhibit either increasing or decreasing returns to scale as their respective computed value of F (=32.7393, 329.4594 and 285.2392 respectively) is greater than the critical value of F (=6.85, 7.08 and 7.08 respectively) at 1% level of significance. However, in the Bishnupur district the selected specification exhibit constant returns to scale as the computed value of F (=2.9646) is smaller than critical value of F (=4.00) even at 5% level of significance. These results at the district level—either increasing or decreasing returns to scale-exhibited in the three districts and constant returns to scale exhibited in Bishnupur district indicates the divergent agricultural environment in the valley areas of the state.

6.8. Conclusion

In this chapter we have examined the impact of agricultural capital formation on agricultural output in Manipur employing Cobb-Douglas production function. The following findings were emerged;

1. At macro level land, labour and gross capital formation play dominant role in the growth and production of agricultural output while technology component (HYVs and chemical fertilizer) plays subordinate role.

2. In Manipur private capital formation is relatively more important in explaining the variation in agricultural output. The output elasticity of private capital formation is 0.64 as against 0.28 for public capital formation.
3. At micro level land and accumulated capital stock of farmers explained nearly 92% of the output variation. The micro finding does substantiate the relative importance of private capital formation in raising agricultural output.
4. The findings regarding returns to scale are mixed both at macro and micro level depending on the variable included at the specification and area of study.

NOTES

1. Chand, Ramesh (2000): *op. cit.*, p. 50.
2. Mundlak, Yair et. al. (2000): *The Determinants of Agricultural Production: A Cross Country Analysis*, World Bank., WPS 1927, p. 20.
3. Sankhayan, P.L. (1986): *Introduction to the Economics of Agricultural Production*, Prentice Hall of India Pvt. Ltd., New Delhi, p. 3.
4. *ibid.*, p. 3.
5. Debertin, David L. (1986): *Agricultural Production Economics*, McMillan Publishing Company, New Delhi, p. 14.
6. Since the beginning of 90's, a number of scholars have questioned on the coverage and definitional aspect of public Capital formation. A survey of literature in this regard is presented in Chapter-IV.
7. Mallick, S.K. (1993): *op. cit.*, pp. 674-675.
8. The expenditure heads included in the broad series has been reproduced in Chapter IV, p. 114.
9. As the database on private investment is based on the decennial survey, AIDIS and DIS of RBI and NSSO respectively only two points could be obtained.

10. Chand, Ramesh (2000): *op. cit.,* pp. 50-57.
11. The budgetary heads of capital expenditure included in the broad series has already been reproduced in Chapter IV, p. 114.
12. Physical variables utilised in the analysis are canal intensity and power supplied to agriculture. While financial variables are the cumulative public investment in agriculture or irrigation based on CSO series, cumulative investment in irrigation plus power and yearly investment in agriculture and allied activities as presented in the budgetary heads of expenditure (broad series).
13. Gulati, Ashok and Seema Bathla (2000): *op. cit,* pp. 57-60.
14. *ibid.,* p. 61.
15. Mundlak, Yair et al. (2000): *op. cit.,* p. 19.
16. Sustainable agricultural development has attracted a number of scholars and researchers in recent years. "Sustainable agriculture is the successful management of resources for agriculture to satisfy the changing human needs while maintaining or enhances the quality of the environment, and conserving the natural resources."
17. Singh, W. Kumar (1985): *op. cit.,* pp. 188-89.
18. Factors such as rainfall, terms of trade, interaction term between terms of trade and technology are also found to have significant impact on agricultural production. (Misra and Hazell, 1996 and Misra, 1998).
19. The estimated growth rate of area, production and yield of rice and the decomposition of growth of rice output in the state have been shown in Chapter II, section 2.3.
20. Mundlak, Yair et at, (2000): *op. cit.,* p. 7.
21. We used the term 'technological basket' because technology involved many ingredients such as HYVs, Fertilizers, Pesticides, Insecticides, mechanical tools, methods and techniques, etc.
22. Singh, W. Kumar (1985): *op. cit.,* p. 171.
23. For more detailed explanation on the various reasons for inclusion of error term 'u_t' in the regression analysis see Gujarati, Damodar N. (1995): *op. cit.,* pp. 39-41.
24. Cyclical movement of the time series data is generally known as sluggishness or inertia. For more detail see *ibid.,* p. 403.

25 Heady, E. O. and John L. Dillon (1961): *Agricultural Production Function*, Kalyani Publishers, Ludhiana, p. 69.

26. The estimates of returns to scale may be biased unless all input factors are included in the production function. Returns to scale will be underestimated if the excluded inputs vary less than proportionately with changes in the included factors over the sample observations. If the reverse situation holds true the elasticity of production or returns to scale will be overestimated. For more detail see *Ibid.*, p. 230-231.

27. Gujarati, Damodar N. (1995): *op. cit.*, pp. 256-257.

28. *ibid.*, pp. 257-258.

7

Policy Implications

Manipur has remained a dominantly agrarian economy. Though the share of agriculture in NSDP has declined, the proportion of people dependent on agriculture remains quite high. Green revolution seems to have bypassed the state. The agrarian structure is such that the dominantly marginal and small farmers are not in a position to take full advantage of the capital intensive HYV technology. While area under rice cultivation registered a declining growth rate during 1972-1999, production grew because of the growth in yield rate. For a three yearly moving average series the entire increase in rice output came from yield effect.

This study for the first time considered capital formation as an explicit factor of agricultural production for Manipur by estimating the time series of capital formation for the period 1972 to 1997. It also studied the determinants of capital formation in agricultural sector of Manipur. Public capital formation and private capital formation are treated separately. Public capital formation is considered as an exogenous variable and an attempt has been made to identify the determinants of private capital formation in terms of public capital formation lagged by one period, terms of trade, technology and interaction between technology and terms of trade. High correlation coefficient implies that private capital formation in agricultural sector of Manipur is affected by these variables. Regression analysis shows that terms of trade have significantly affected the private capital formation in agriculture. The estimated elasticity of private capital formation with respect to terms of trade is -0.612. The negative sign on estimated coefficient of terms of trade can be interpreted as follows.

The negative sign implies that as terms of trade improves private capital formation in agriculture declines. It may be due to the fact that farmers have an explicit monetary target to achieve. Once terms of trade improve for the farmer he gets more revenue from his produce. Once the increase in revenue reaches the target there is no incentive for him to invest more in agriculture which is not considered a profitable venture. The setting of the low monetary targets may be due to ignorance of alternative investment opportunities by the farmers. This is associated with illiteracy and poverty of the farmers. It implies that once the investment horizon of the farmers is expanded terms of trade may start having the expected impact on private capital formation. Thus, private capital formation can be enhanced by raising remunerative farm prices and the awareness level of the farmers.

It has been noted that the elasticity of private capital formation with respect to public capital formation lagged by one year is 0.091. The positive sign on estimated elasticity implies that public capital formation lagged by one year does stimulate private capital formation. That is to say that complementarity hypothesis between public capital formation and private capital formation does prevail in the agricultural sector of Manipur as well. Private capital formation is expected to follow public capital formation. If the government takes care of the infrastructure by way of public capital formation private capital formation is likely to follow it. It calls for the bigger scope of participation by the government in the agricultural sector.

At macro level analysis it was found that land, labour and capital formation play an important role in the growth and production of agricultural output while technological factors such as diffusion of high yielding varieties and consumption of chemical fertilizers play subordinate role. The output elasticity of land, labour and capital formation is 0.492, 0.364 and 0.346 respectively. Though the estimated output elasticity of land is the largest the declining growth rate of area under rice cultivation registered during 1972-1999 reveals that exploitation of land for rice cultivation seems to have reached the saturation point. In such a situation if the annual growth rate of rice production of 1.51 percent registered during the last three decades continues for another five years then the target set to achieve at the end of Tenth Plan is unlikely to be realized. If we are to achieve the Tenth Plan target of 531 thousand MT the rice production in the state should grow at the rate of 4.99 percent per annum. With the overwhelming dependence of labour on agriculture we seem to experience the classic case of disguised unemployment. More labour can no longer be expected to generate higher agricultural output. In such a situation the only means to fill the gap between the growth rate experience during the last three decades and required rate of growth is to step up the cropping intensity by adopting capital intensive technique of production. This calls for a larger amount of agricultural capital formation. If Manipur wants to feed her people, higher capital formation in agricultural sector is a must.

Our findings show that at the macro level private capital formation is relatively more important in explaining the variation in agricultural output. The output elasticity of private capital formation is 0.64 against 0.276 for public capital formation. At the micro level based on the sample of 294 farmers, land and accumulated capital stock of farmers explained nearly 92 percent of the output variation. More land and accumulated capital stock of farmers will lead to higher output. Intensification of cropping intensity will have the same effect as expansion of land. This will lead to more productive utilization of labour currently passing through the phase of disguised unemployment. The micro findings thus substantiate the relative importance of private capital formation in raising agricultural output.

The findings regarding constant returns to scale are mixed both at macro and micro level depending on the variables included in the specification and area of study.

Manipur used to be self-sufficient in rice production leading to some amount of export. However, with the integration of state economy into the mainland economy agriculture began to be neglected. The investment flowing into this sector was not commensurate with the requirements of the sector. It was also not supported by a proper plan for this sector. As a result the benefits of green revolution bypassed this state.

An analysis of the trend growth rate shows that agricultural output has been growing at a constant rate. There was no acceleration in the growth rate as a result of change in agricultural technology embodied in HYV technology. The irrigation infrastructure created remained under utilized and in name only. The package concept of HYV technology was misunderstood. An agricultural sector whose structure is dominated by small and marginal farmers can not be expected to capitalise on a capital intensive technology. As a result today the state is a net importer of food grains. Under these circumstances self-sufficiency in food grains is a far cry.

The present study outlines a strategy for growth of agricultural output by bringing in capital formation as an explicit factor of production. Public capital formation should first create an environment for the farmers. The farmers should be made aware of alternative investment opportunities through intensive extension networks. They should be made aware of the appropriate technologies available at the same time. They should be advised on how to invest their money. This should be supplemented with adequate provision for remunerative price for the farmers. This will improve their terms of trade and private capital formation will follow.

In Manipur there is practically no scope for raising the area under cultivation, which registered a declining growth rate in the last three decades. Therefore, the important role of capital formation particularly private capital formation cannot be ignored. It also provides a way for raising agricultural output when we have already reached the saturation point for land. Regarding labour the overwhelming dependence of workers on agriculture may have already created the classic case of disguised unemployment. More labour can no longer be expected to generate higher agricultural output. The development of agricultural sector has wide ranging implications for a small and backward state like Manipur. If agriculture prospers all other sector will show signs of improvement. It will also have implications for our efforts towards eradication of poverty.

Appendices

APPENDIX 2.1

Trends of Area Under Paddy by Different Types of Cultivation

(in '000 hectares)

Year	*Jhum*	*Permanent Terrace*		*Total*
		Valley districts	*Hill districts*	
1985-86	25.23	98.15	41.22	164.60
1986-87	25.64	101.84	37.66	165.14
1987-88	25.50	105.45	34.81	165.76
1988-89	26.44	99.10	41.52	167.06
1989-90	26.57	99.73	36.07	162.37
1990-91	23.81	101.34	32.26	157.41
1991-92	24.29	98.89	37.12	160.30
1992-93	25.24	87.39	38.56	151.19
1993-94	26.57	93.02	42.28	161.87
1994-95	37.77	96.35	29.81	163.93
1995-96	29.79	77.52	27.62	134.93
1996-97	43.48	88.95	33.67	166.10
1997-98	40.45	88.42	29.03	157.90
1998-99	36.90	97.93	32.06	166.89
1999-2000	40.54	89.24	27.27	157.05

Source: Crop Estimation Survey (Annual), Department of Economics and Statistics, Government of Manipur.

APPENDIX 2.2A

Requirement of Mandays for Cultivation of 1 (one) ha. of Paddy Under Transplantation (HYV)

Sl. No.		*Particulars*	*Mandays*
A.	1.	Ploughing 6 times	36
	2.	Repairing & making of bunds	8
	3.	Puddling/planking 4 times	8
	4.	Preparation of nursery bed and fertilizers application	5
	5.	Application of fertilizers (main field)	4
	6.	Sowing and nursery management	6
	7.	Uprooting and distributions of seeding for Transplantation	12
	8.	Transplanting in lines	40
	9.	Weeding 2 (two) times	48
	10.	Top dressing of fertilizers 2 (two) times	8
	11.	Plant protection measures 2 (two) times	8
	12.	Clearing of irrigation cannels and water management	8
	13.	Harvesting	32
	14.	Threshing, winnowing, drying and bagging	48
		Total =	271

i. Cost of labour @ Rs. 44.65 per head/day (Valley) = Rs. 12,100.15
ii. Cost of labour @ Rs. 47.65 per head/day (Hills) = Rs. 12,913.15

B. Fertilisers

1. Urea 100 kg @ Rs. 3.32/kg = Rs. 332.00
2. DAP 100 kg @ Rs. 10.45/kg = Rs. 1,045.00
3. MOP 50 kg @ Rs. 5.22/kg = Rs. 261.00

Total = Rs. 1,638.00

C. Seed 60 kg @ Rs. 6.50/kg = Rs. 390.00

D. P.P. Chemicals = Rs. 400.00

Total cost of production (A+B+C+D) = Rs. 14,528.15 (Valley)
Rs. 15,341.15 (Hills)

Yield per hectare = 5 tonnes

Source: Directorate of Agriculture, Government of Manipur.

APPENDIX 2.2B

Requirement of Mandays for Cultivation of 1 (one) ha. of Paddy Under Transplantation (Local)

Sl. No.		*Particulars*	*Mandays*
A.	1.	Ploughing 6 times	30
	2.	Repairing & making of bunds	8
	3.	Puddling/planking 4 times	8
	4.	Preparation of nursery bed and fertilizers application	5
	5.	Application of fertilizers (main field)	4
	6.	Sowing and nursery management	6
	7.	Uprooting and distributions of seeding for Transplantation	10
	8.	Transplanting in lines	40
	8.	Weeding 2 (two) times	44
	9.	Top dressing of fertilizers 2 (two) times	4
	10.	Plant protection measures 2 (two) times	8
	11.	Clearing of irrigation cannels and water management	8
	12.	Harvesting	28
	13.	Threshing, winnowing, drying and bagging	42
		Total =	247

i. Cost of labour @ Rs. 44.65 per head/day (Valley) = Rs. 11028.55
ii. Cost of labour @ Rs. 47.65 per head/day (Hills) = Rs. 11769.55

B. **Fertilisers**

4. Urea 50 kg @ Rs. 3.32/kg = Rs. 166.00
5. DAP 50 kg @ Rs. 10.45/kg = Rs. 522.50
6. MOP 50 kg @ Rs. 5.22/kg = Rs. 130.50

Total = Rs. 819.00

C. Seed 60 kg @ Rs. 6.50/kg = Rs. 390.00

D. P.P. Chemicals = Rs. 300.00

Total cost of production (A+B+C+D) = Rs. 12,537.55 (Valley)
Rs. 13,278.55 (Hills)

Yield per hectare = 3 tonnes

Source: Directorate of Agriculture, Government of Manipur.

APPENDIX 2.3

Trend of Area, Production and Yield of Rice Cultivation in Manipur

Year	*Area (in '000 ha.)*	*Production (in '000 tonnes)*	*Yield (in kg/ha.)*
1972-73	146.60	152.16	1037.93
1973-74	169.78	238.62	1405.47
1974-75	176.51	274.68	1556.17
1975-76	177.19	276.42	1560.02
1976-77	177.20	266.97	1506.60
1977-78	179.13	299.98	1674.65
1978-79	159.56	255.21	1599.46
1979-80	155.20	227.53	1466.04
1980-81	188.55	273.04	1448.10
1981-82	168.30	253.14	1504.10
1982-83	158.49	219.47	1384.76
1983-84	161.22	255.13	1582.50
1984-85	167.41	332.99	1989.07
1985-86	164.60	332.53	2020.23
1986-87	165.14	242.47	1467.27
1987-88	165.76	312.76	1886.82
1988-89	167.07	319.72	1913.80
1989-90	162.37	245.12	1509.64
1990-91	157.41	274.17	1741.76
1991-92	160.30	340.45	2123.83
1992-93	151.19	273.12	1806.47
1993-94	161.87	348.82	2154.94
1994-95	163.93	358.51	2186.97
1995-96	134.93	338.05	2505.37
1996-97	166.10	367.28	2211.20
1997-98	157.90	351.67	2227.17
1998-99	166.89	382.19	2290.07
1999-2000	157.05	364.98	2323.97

Source: Various issues of Crop Estimation Survey of Manipur Department of Economics and Statistics, Govt. of Manipur

APPENDIX 5.1

Trends of Terms of Trade Index, Percentage of Area Under HYV of Rice Crop and the Interaction Term between the Two

Year	*TT*	*TECH*	*INT*
1972-73	110.26	4.775	607.13
1973-74	116.69	7.068	808.79
1974-75	109.18	16.996	1843.7
1975-76	104.41	19.753	2066.5
1976-77	96.61	22.573	2260.0
1977-78	90.05	25.121	2137.6
1978-79	90.54	30.083	2555.2
1979-80	95.51	25.773	2618.3
1980-81	103.72	27.049	2704.9
1981-82	107.45	39.216	4297.3
1982-83	115.48	45.429	5123.0
1983-84	121.04	47.761	5926.2
1984-85	121.27	47.787	6033.6
1985-86	122.97	48.603	5514.9
1986-87	127.09	50.260	6419.1
1987-88	124.04	48.263	6690.6
1988-89	122.48	48.486	5058.5
1989-90	114.63	55.429	6899.8
1990-91	118.93	57.176	6580.3
1991-92	125.19	57.392	6727.5
1992-93	128.63	46.299	6633.3
1993-94	130.63	44.480	5577.4
1994-95	124.06	43.921	5412.4
1995-96	123.98	51.879	6410.1
1996-97	124.62	43.347	5424.9
1997-98	125.14	45.598	5706.2

Note: 1. TT=Terms of trade index have been constructed based on the data provided in "National Account Statistics of India: 1950-51 to 1996-97", Economic and Political Weekly Research Foundation, Mumbai.

b. "Estimates of Net State Domestic Product (NSDP)", Directorate of Economics and Statistics, Government of Manipur.

2. Percentage of Area under HYV Paddy which represent Technology (TECH) variable have been arrived from the time series of Area under HYV Paddy provided in various "Fertiliser Statistics", Fertiliser Association of India, New Delhi.

3. Interaction term between the Terms of Trade and Technology is arrived by multiplying Terms of Trade Index and Percentage of Area under HYV Paddy.

APPENDIX 6.1(A)

Trends of Area Under HYV Paddy and Fertiliser Consumption in the State

Year	*Area under HYV Paddy ('000 ha.)*	*Fertiliser Consumption ('000 tonnes)*
1972-73	7	1.31
1973-74	12	1.55
1974-75	30	1.81
1975-76	35	1.74
1976-77	40	2.31
1977-78	45	2.47
1978-79	48	2.60
1979-80	40	3.01
1980-81	51	2.83
1981-82	66	3.33
1982-83	72	4.02
1983-84	77	4.34
1984-85	80	3.76
1985-86	80	4.66
1986-87	83	5.66
1987-88	80	6.33
1988-89	81	7.30
1989-90	90	6.90
1990-91	90	8.22
1991-92	92	8.83
1992-93	70	9.16
1993-94	72	9.11
1994-95	72	8.55
1995-96	70	11.82
1996-97	72	12.18
1997-98	72 *	13.19

Source: 1. "Fertiliser Association Statistics", Fertiliser Association of India (various issues).

2. Statistical Abstract of Manipur 2002, Department of Economics and Statistics, Govt. of Manipur.

Note: * Assume as that of 1996-97 (Data for 1997-98 is yet to be published.

APPENDIX 6.1(B)

Trends of Area Under Rice Crop and Agricultural Labourer

Year	*Area under Rice Crop('000 ha.)*	*Agricultural Labourer ('000 tonnes)*
1972-73	146.60	272.90
1973-74	169.78	284.20
1974-75	176.51	295.90
1975-76	177.19	308.20
1976-77	177.20	321.00
1977-78	179.13	334.30
1978-79	159.56	348.10
1979-80	155.20	362.60
1980-81	188.55	377.60
1981-82	168.30	393.20
1982-83	158.49	401.60
1983-84	161.22	410.10
1984-85	167.41	418.70
1985-86	164.60	427.60
1986-87	165.14	436.60
1987-88	165.76	445.90
1988-89	167.06	455.30
1989-90	162.37	465.00
1990-91	157.41	474.80
1991-92	160.30	484.80
1992-93	151.19	495.10
1993-94	161.87	505.60
1994-95	163.93	516.30
1995-96	134.93	527.20
1996-97	166.10	538.40
1997-98	157.90	549.80

Source: 1. Various issues of crop estimation survey, Department of Economics and Statistics, Govt. of Manipur.

2. Census of India 1971, 1981 and 1991, Manipur.

Bibliography

Abel, Andrew B. (1980): "Empirical Investment Equations: An Integrative Framework", Karl Brunner & Allen H. Meltzer (eds.): *On the State of Macro-Economics*, Amsterdam.

Agarwala, A.N. & S.P. Singh (1971): *The Economics of Underdevelopment*, Reprint, Oxford University Press, New York.

Agrawal, A.N. (1998): *Indian Economy: Problems of Development and Planning*, 24th Edition, Wishwa Prakashan, New Delhi.

Ahuja, Kanta (1986): "Agricultural Labour and Rural Employment", M.L. Dantwala & Others (eds.): *Indian Agricultural Development since Independence: A Collection of Essays*, Oxford & IBH Publishing Co. Pvt. Ltd., New Delhi, pp. 400-431.

Ahluwalia, I.J. & C. Rangarajan (1986): *A Study of Linkages between Agriculture and Industry: The Indian Experience*, Proceedings of the World Economic Congress of IEA, New Delhi.

Alagh, Yoginder K. (1997): "Agricultural Investment and Growth", *Indian Journal of Agricultural Economics*, vol. 52, no. 2, April-June, pp. 279-287.

Alamgir, Mohiuddin (1976): "Rural Savings and Investment in Developing Countries: Some Conceptual and

Empirical Issues", *Bangladesh Development Studies*, vol. 4, no. 1, pp. 1-47.

Aschauer, David Alan (1989): Does Public Capital Crowd Out Private Capital?, *Journal of Monetary Economics*, vol. 24, no. 2, September, pp. 171-188.

Banerjee, Amalesh (1996): "Dynamic Capital Formation in Agriculture and Financial Reform", *Indian Journal of Agricultural Economics*, vol. 51, no. 4, Oct.-Dec., pp. 560-568.

Bansil, P.C. (1984): *Agricultural Statistics in India,* 3rd Edition, Oxford & IBH Publishing Co. Pvt. Ltd., New Delhi.

Barker, R., Robert W. Herdt & Beth Rose (1985): "The Rice Economy of Asia", *Resource for the Future*, Washington D.C.

Bhalla, G.S. (1977): *Changing Agrarian Structure in India: A Study of the Impact of Green Revolution in Haryana*, Meenakshi Prakashan, Delhi.

Bhalla, G.S. & Peter B.R. Hazell (1997): "Food grains Demand in India to 2020: A Preliminary Exercise", *Economic and Political Weekly*, vol. XXXII, no. 52, Dec. 27, pp. A-150-A-154.

Bhalla, Sheila (1987): "Trends in Employment in Indian Agriculture, Land and Asset Distribution", *Indian Journal of Agricultural Economics*, vol. 42, no. 4, Oct.-Dec., pp. 537-560.

Bhatia, H.L. (1993): *Public Finance*, 17th Edition, Vikas Publishing House Pvt. Ltd., New Delhi.

Bhatia, M.S. (1999): "Rural Infrastructure and Growth in Agriculture", *Economic and Political Weekly*, vol. XXXIV, no. 13, March 27-April 2, pp. A-43-A-48.

Blejer, Mario I. & Mohsin S. Khan (1984): *Government Policy and Private Investment in Developing Countries*, IMF Staff Papers 31, pp. 379-403.

Bohm-Bawerk, Eugen V. (1948): *The Positive Theory of Capital* trans. by William Smart, G.E. Stechert and Co., New York.

Booth, Anne & R.M. Sundrum (1984): *Labour Absorption in Agriculture: Theoretical Analysis and Empirical Investigations*, Oxford University Press, Delhi.

Central Statistical Organisation (1969): "Estimates of Capital Formation in India 1960-61 to 1965-66", *National Income Statistics (Preliminary)*, Department of Statistics, Government of India.

------: (1989): *National Accounts Statistics: Sources and Methods*, Department of Statistics, Ministry of Planning, Government of India, New Delhi.

------: *National Accounts Statistics*, (Various Issues).

Chakravarty, Sukhamoy (1980): *Alternatives in Agricultural Development*, ICSSR Working Group Report (Allied), New Delhi.

Champernowne, D.G. (1953-54): "The Production Function and The Theory of Capital: A Comment", *The Review of Economic Studies*, vol. 21(2), no. 55, pp. 112-135.

Chand, Ramesh (2000): *Emerging Trends and Regional Variations in Agricultural Investments and their Implication for Growth and Equity*, Policy Paper-11, National Centre for Agricultural Economics and Policy Research, New Delhi.

------: (2001): "Emerging Trends and Issues in Pubic and Private Investments in Indian Agriculture: A state-wise Analysis", *Indian Journal of Agricultural Economics*, vol. 56, no. 2, April-June, pp. 161-184.

Chand, Ramesh & Parmod Kumar (2004): Determinants of Capital Formation and Agricultural Growth: Some New Explanations, *Economic and Political Weekly*, vol. XXXIX, no. 52, Dec. 25, pp. 5611-5616.

Chattopadhyay, Manabendu & Atanu Sengupta (1997): "Farm Size and Productivity A New Look at the Old Debate", *Economic and Political Weekly*, vol. XXXII, no. 52, Dec. 27, pp. A-172 - A-176.

Chaturvedi, D.N. & Amaresh Bagchi (1984): "Age Structure of Fixed Capital in the Indian Economy 1970-71 & 1980-81", *The Journal of Income and Wealth*, vol. 7, no. 1, pp. 82-87.

Clark, Colin (1957): *The Conditions of Economic Progress*, 3rd Edition, MacMillan & Co. Ltd., London.

Crego, Al, Donald Larson, Rita Butzer & Yair Mundlak (1998): *A New Database on Investment and Capital for Agriculture and Manufacturing*, Policy Research Working Paper 2013, http://eco.worldbank.org/external/default/main?

Dandekar, V.M. (1986): *Agriculture, Employment and Poverty*, Seminar Paper presented at the Conference Organised by the Centre for Asian Development Studies, Boston University, October.

Dantwala, M.L. (1986): "Strategy of Agricultural Development since Independence", M.L. Dantwala & Others (eds.): *Indian Agricultural Development since Independence: A Collection of Essays*, Oxford & IBH Publishing Co. Pvt. Ltd., New Delhi pp. 1-15.

M.L. Dantwala & Others (eds.): *Indian Agricultural Development since Independence: A Collection of Essays*, Oxford & IBH Publishing Co. Pvt. Ltd., New Delhi pp. 59-88.

Deb, Surajit (2004): Terms of Trade and Investment Behaviour in Indian Agriculture: A Cointegration Analysis, *Indian Journal of Agricultural Economics*, vol. 59, no. 2, April-June, pp. 209-230.

Debertin, David L. (1986): *Agricultural Production Economics*, MacMillan Publishing Company, New York.

Desai, Bhupat M. (1994): "Contributions of Institutional Credit, Self-Finance and Technological Change to Agricultural Growth in India", *Indian Journal of Agricultural Economics*, vol. 49, no. 3, July-Sept., pp. 457-475.

Desai, Bhupat M. & Errol D'Souza (1999): "Economic Reforms, Terms of Trade, Aggregate Supply and Private Investment in Agriculture, *Economic and Political Weekly*, vol. XXXIV, no. 20, May 15-21, pp. 1220-1224.

Desai, Bhupat M. & N.V. Namboodri (1997): "Strategy and Sources of Growth in Crop Agriculture", Bhupat M. Desai (ed.): *Agricultural Development Paradigm for the Ninth Plan under New Economic Environment*, CMA Monograph no. 179, Oxford & IBH Publishing Co. Pvt. Ltd., New Delhi, pp. 205-231.

------: "Government Expenditure on Agriculture under Planning Era", Bhupat M. Desai, (ed.): *Agricultural Development Paradigm for the Ninth Plan under New Economic Environment*, CMA Monograph no.

179, Oxford & IBH Publishing Co. Pvt. Ltd., New Delhi pp. 979-995.

------ (1997): "Determinants of Total Factor Productivity in Indian Agriculture", *Economic and Political Weekly*, vol. XXXII, no. 52, Dec. 27, pp. A-165-A-171.

Deshpande, R.S. (1998): "Land Reforms and Agrarian Structure in Maharashtra", *Journal of Indian School of Political Economy*, vol. X, no. 1, Jan.- March, pp. 1- 23.

Dharam Narain (1976): *Growth of Productivity in Indian Agriculture*, Occasional Paper no. 93, Cornell University.

Dhawan, B.D. (1996a): "Relationship between Public and Private Investments in Indian Agriculture with Special Reference to Public Canals", *Indian Journal of Agricultural Economics*, vol. 51, nos. 1& 2, Jan-June, pp. 209- 219.

------ (1996b): "Price and Non-Price Factors in Agricultural Investments"(Discussion), *Economic and Political Weekly*, vol. XXXI, no. 25, June 22, p. 1632.

------ (1996c): "Trends and Determinants of Capital Investments in Agriculture", *Indian Journal of Agricultural Economics*, vol. 51, no. 4, Oct.-Dec., pp. 529- 542.

------ (1997a): "Price and Non-Price Factors in Agricultural Investments" (Discussion), *Economic and Political Weekly*, vol. XXXII, no. 5, Feb.1, pp. 235- 236.

------ (1997b): "Production Benefits from Large Scale Canal Irrigation", *Economic and Political Weekly*, vol. XXXXII, no. 52, Dec. 27, pp. A-177-A-181.

------ (1998): *Studies in Agricultural Investments and Rural Savings*, Commonwealth Publishers, New Delhi.

Dhawan, B.D. & S.S. Yadav (1995): "Private Fixed Capital Formation in Agriculture: Some Aspects of Indian Farmers' Investment Behaviour", *Economics and Political Weekly*, vol. XXX, no. 39, Sept. 30, pp. A-103-A-109.

------ (1997): "Public Investment in Indian Agriculture: Trends and Determinants", *Economic and Political Weekly*, vol. XXXII, no. 14, April-5, pp. 710-714.

Dholakia, Ravindra & Bakul H. Dholakia (1993): "Growth of Total Factor Productivity in Indian Agriculture", *Indian Economic Review*, vol. XXVIII, no. 1, Jan.-June, pp. 25-40.

Economic and Political weekly Research Foundation (1998): *National Accounts Statistics of India: 1950-51 to 1996-92*, 3rd Edition, Mumbai, October.

Gandhi, Vasant P. (1990): "Investment Behabiour in Developing Countries: The Case of Agriculture in India", *Food Research Institute Studies* vol. XXII, no. 1.

------ (1996): "Investment Behaviour in Indian Agriculture", *Indian Journal of Agricultural Economics*, vol. 51, no. 4, Oct.- Dec., pp. 543-559.

------ (1997): "The Role of Agriculture in India's Economic Development: A Survey", Bhupat M. Desai (ed.): *Agricultural Development Paradigm for the Ninth Plan Under New Economic Environment*, CMA Monograph no. 179, Oxford and IBH Publishing Co. Pvt. Ltd., New Delhi, pp. 108-121.

Gandhi, Vasant P. & Gyanendra Mani (1997): *Investment Behaviour in Indian Agriculture: A Theoretical and Empirical Investigation*, Indian Institute of Management, Ahmedabad (Mimeo).

Ghosh, A. & K. Dutt (1977): *Development of Capitalist Relations in Agriculture (A Case Study of West Bengal 1993- 1971)*, People's Publishing House, New Delhi.

Ghosh, A. (2005): *Capital Formation and Entrepreneurship in Indian Agriculture*, Concepts Publishing Company, New delhi.

Governmetn of Andha Pradesh (1996): *Estimates of Gross Fixed Capital Formation in Andhra Pradesh: 1989-90 to 1991- 92*, Directorate of Economics and Statistics, Hyderabad.

Government of India (1953): *First Five Year Plan*, Planning Commission, New Delhi.

------ (1983): *Agricultural Census of India: 1976-77*, Department of Agriculture & Cooperation Ministry of Agriculture, New Delhi.

------ (1987): "Sarvekshana", *Journal of the National Sample Survey Organization*, Department of Statistics, Ministry of Planning, vol. XI, no. 2, Issue no. 33, Oct.

------ (1988): "A Note on Some Aspects of Operational Holdings", *Sarvekshana*, July.

------ (1995): *Indian Agriculture in Brief*, 26th Edition, Directorate of Economics and Statistics, Department of Agriculture and Cooperation, Ministry of Agriculture, New Delhi.

------ (1999 & 2000): *Agricultural Statistics: At a Glance 1999 & 2000*, Directorate of Economics and Statistics, Department of Agriculture & Cooperation, Ministry of Agriculture, New Delhi.

------: *Census of India 1971, 1981 & 1991*, Series-15, Manipur

Part iii - B Series, Directorate of Census Operations Manipur.

------ (2001): *Census of India 2001*, Series 14, Manipur Population Totals Paper- 1 & 2 of 2001, Directorate of Census Operations Manipur.

------ (2001): *Agricultural Development in Eastern and North Eastern India*, TFYP Working Group Report Sr. no. 40/2001, Planning Commission.

------ (2003): *Report of the Committee on capital Formation in Agriculture*, Directorate of Economics and Statistics, Department of Agriculture & Cooperation, Ministry of Agriculture, New Delhi.

Government of Karnataka (1994): *Estimates of Gross Capital Formation in Public Sector in Karnataka: 1985- 86 to 1991-92 at Current Prices*, Directorate of Economics and Statistics, Bangalore.

Government of Madhya Pradesh (1996): *Estimates of Gross Fixed Capital Formation in Madhya Pradesh*, Directorate of Economics and Statistics, Bhopal.

Government of Manipur (1925): *Proceeding of Manipur State Darbar 1925*, Unpublished Record in the Secretariat Library, Imphal.

------ (1978 & 1999): *Statistical Handbook of Manipur*, Directorate of Economics and Statistics, Imphal.

------: *Agricultural Census of Manipur: 1970-71, 1976-77, 1980-81, 1985-86, & 1990-91*, Department of Agriculture, Imphal.

------: *Input Survey: 1981-82, 1986-87 & 1991-92*, Department of Agriculture. Imphal.

------ (1998): *Statistical Abstract of Manipur*, Directorate of Economics & Statistics, Imphal.

------: *Budget Documents of State Government* (various issues).

------ (1999): *Annual Plan 1999-2000*, vol. I, Department of Planning, Imphal.

------: *Crop Estimation Survey of Manipur* (various issues), Directorate of Economics and Statistics, Imphal.

------ (2001): *Statistial Abstract of Manipur*, Directorate of Economics and Statistics, Imphal.

Government of Tamil Nadu (1997): *Report on Gross Fixed Capital Formation of Tamil Nadu: 1989-90 to 1991-92*, Directorate of Economics and Statistics, Chennai.

Gujarati, Damodar N. (1995): *Basic Econometrics*, 3rd Edition, McGraw-Hill International Editions, Singapore.

Gulati, Ashok & Anil Sharma (1997): "Freeing Trade in Agriculture Implications for Resource use Efficiency and Cropping Pattern Changes", *Economic and Political Weekly*, vol. XXXII, no. 52, Dec. 27, pp. A-155-A-164.

Gulati, A. & Seema Bathla (2000): *Capital Formation in Indian Agriculture: Trends, Composition and Implications for Growth*, Mimeo, Institute of Economic Growth, Delhi.

------ (2001): "Capital Formation in Indian Agriculture: Revisiting the Debate", *Economic and Political Weekly,* vol. XXXI, no. 20, May 19, pp. 1697-1708.

Heady, E.O. (1950): *Capital Formation in Agriculture*, Mimeo, Chicago.

------ (1971): *Economic Models and Quantitative Methods for Decisions and Planning in Agriculture*, The Iowa State University Press, AMES, Iowa, USA.

Heady, E.O. & John L. Dillon (1961): *Agricultural Production Functions*, Kalyani Publishers, Ludhiana.

Hicks, J.R. (1946): *Value and Capital: An Inquiry into Some Fundamental Principles of Economic Theory*", 2nd Edition, Oxford University Press, London.

Hirashima, S. (2000): Issues in Agricultural Reforms: Public Investment and Land Market Development, *Economic and Political Weekly*, vol. XXXV, Nos. 43 & 44, Oct. 21 - Nov. 3, pp. 3879-3884.

Indian Council of Agricultural Research (1986): *Annual Report*, ICAR Complex for NEH Region, Manipur Centre, Babupara.

------ (1992): *Soils of Manipur for Land use Planning*, National Bureau of Soil Survey and Land use Planning, Regional Centre, Jorhat.

Islamuddin, Mohd. (1984): *Problems and Prospects of Agricultural Marketing in Manipur during the Plan Period (1951-52 to 1984-85)*, Unpublished Ph. D. Thesis, Manipur University.

Jain, Sharad Chandra (1968): *Price Behaviour and Resource Allocation in Indian Agriculture*, Allied Publishers Private Ltd., Bombay.

Jairath, M.S. & Brijesh C. Purohit (1996): "Trends in Capital Formation in Agriculture: A Case of Arid India", *Indian Journal of Agricultural Economics*, vol. 51, no. 4, Oct.-Dec., pp. 587-596.

Jhingan, M.L. (1982): *The Economics of Development and Planning (with special reference to India)*, 15th Edition, Vikas Publishing House Pvt. Ltd., New Delhi.

Jorgenson, D.W. (1971): "Econometric Studies of Investment

Behaviour: A Survey", *Journal of Economic Literature*, vol. 9, no. 4, December, pp. 1111-1147.

Just, Richard E., David Zilberman, & Eithan Hochman (1983): "Estimation of Multicrop Production Functions", *American Journal of Agricultural Economics*, vol. 65, November, pp. 770-780.

Kahlon, A.S. & Karam Singh (1984): *Managing Agricultural Finance*, Allied Publishers Pvt. Ltd., New Delhi.

Kalirajan, K.P. & R.T. Shand (1997): "Sources of Output Growth in Indian Agriculture", *Indian Journal of Agricultural Economics*, vol. 52, no. 4, Oct.-Dec., pp. 693-706.

Kalra, O.P. (1973): *Agricultural Policy in India*, Popular Prakashan, Bombay.

Karmakar, K.G. (1998): "Capital Formation in Agriculture and Role of Institutional Credit", *Journal of Rural Development*, vol. 17(1), pp. 37-52, NIRD, Hyderabad.

Karunakaran, K.R.& K. Palanisami (1998): "An Analysis of Impact of Irrigation on Cropping Intensity in Tamil Nadu", *Indian Economic Review*, vol. XXXIII, no. 2, pp. 207-220.

Kumar, A. Ganesh (1993): "Falling Agricultural investment and its Consequences", *Economic and Political Weekly*, vol. 27, no. 42. October 17, pp. 2307-2312.

Kumar, Baldev (1969): "Capital Formation in Indian Agriculture", *Indian Journal of Agricultural Economics*, vol. 24, no. 4, April-June, pp. 13-17.

Kurian, A. P. (1969): "Estimates of Private Capital Expenditure in Agriculture During the Period 1969-70 to 1973-74", *Indian Journal of Agricultural Economics*, vol. 24, no. 4, Oct-Dec., pp. 67-71.

Kuznets, Simon (1969): *Economic Growth and Structure*.

Selected Essays, Oxford & IBH Publishing Co. Pvt. Ltd., New Delhi.

Lal, Ram N. (1980): *Capital Formation and its Financing in India*, Monograph in Economics no. 13, Delhi School of Economics, Allied Publishers Pvt. Ltd., New Delhi.

Larson, Donald F., Rita Butzer, Yair Mundlak and Al Crego (2000): A Cross-Country Database for Sector Investment and Capital, *The World Bank Economic Review*, vol. 14, no. 2, pp. 371-391, http://www.worldbank.org/research/journals/wber/may00/may00.htm.

Laxminarayan, H. & S.S. Tyagi (1982): *Changes in Agrarian Structure in India*, Agricole Publishing Academy, New Delhi.

Levi, John (1979): "Traditional Agricultural Capital Formation", *World Development (Incorporating New Commonwealth)*, vol. 7, no. 11/12, Nov./Dec., pp. 1053-1062, Pergamon Press Ltd., Great Britain.

Lewis, W. Arthur (1971): "Economics Development with Unlimited Supplies of Labour", A.N. Agarwala & S.P. Singh (eds.): *The Economics of Underdevelopment*, Oxford University Press, Calcutta, pp. 400-449.

Lutz, F.A. (1961): "The Essentials of Capital Theory", F.A. Lutz & D.C. Hague (eds.): *Theory of Capital*, News York.

Mallick, Sushanta Kumar (1993): "Capital Formation in Indian Agriculture: Recent Trends", *Indian Journal of Agricultural Economics*, vol. 48, no. 4, Oct.-Dec., pp. 667-677.

Mander, G. S. & S.S. Grewal (1988): "Economic Rationale

of Investments on Punjab Farms with Special Reference to Tractorisation", *Indian Journal of Agricultural Economics*, vol. 43, no. 4, Oct.-Dec., pp. 613-623.

Mani, K.P., P. Shaheena & P. Chacko Jose (1996): "Some Reflections on Capital Formation in Indian Agriculture", *Indian Journal of Agricultural Economics*, vol. 51, no. 4, Oct.-Dec., pp. 569-578.

Meier, Gerald M. (1970): *Leading Issues in Economic Development*, Oxford University Press, New Delhi.

Minhas, B.S. & A. Vaidyanathan (1965): "Growth of Crop Output in India, 1951-54 to 1958-61: An Analysis by Component Elements", *Journal of Indian Society of Agricultural Statistics*, vol. XVII, no. 2.

Mishra, S.N. (1996): "Capital Formation and Accumulation in Indian Agriculture since Independence", *Indian Journal of Agricultural Economics*, vol. 51, nos. 1 & 2, Jan-June, pp. 193-208.

Mishra, S.N. & Ramesh Chand (1995): "Public and Private Capital formation in Indian Agriculture: Comments on the Complimentarily Hypothesis and Others", *Economic and Political Weekly*, vol. 30, no. 25 June 24, pp. A-64-A-79.

Misra, V.N. (1998): "Economic Reforms, Terms of Trade, Aggregate Supply and Private Investment in Agriculture: Indian Experience", *Economic and Political Weekly*, vol. XXXIII, no. 31, Aug. 1-7, pp. 2105-2109.

Misra, V.N. & Peter B.R. Hazell (1996): "Price and Non-Price Factors in Agricultural Investments", *Economic and Political Weekly*, vol. XXXI, no. 43, Oct. 26, pp. 2891-2892.

------ (1997): "Price and Non-Price Factors in Agriculture Investments", *Economic and Political Weekly*, vol. XXXII, no. 31, Aug 2-8, pp. 1991-1992.

Mitra, Ashok K. (1997): "Public and Private Investments in Agriculture", Bhupat M. Desai (ed.): *Agricultural Development Paradigm for the Ninth Plan under New Economic Environment*, CMA Monograph no. 179, Oxford & IBH Publishing Co. Pvt. Ltd., New Delhi, pp. 941-978.

Mohanty, B. B. (1999): "Agricultural Modernisation and Social inequality: Case Study of Satara District", *Economic and Political Weekly*, vol. XXXIV, no. 26, June 26, pp. A-50-A-61.

Mujumdar, N.A. & K.A. Menon (1986): "Saving and Capital Formation in the Agricultural Sector: A Review", M.L. Dantwala & Others (eds.): *Indian Agricultural Development since Independence: A collection of Essays*, Oxford & IBH Publishing Co. Pvt. Ltd., New Delhi, pp. 234-269.

Mundlak, Yair, Donal F. Larson & Rita Butzer (2000): *The Determinants of Agricultural Production: A Cross-Country Analysis*, World Bank, Washington D.C.

Murgai, Rinku (2000): *The Green Revolution and the Productivity Paradox: Evidence from the Indian Punjab*, Development Research Group, World Bank, Washington D.C.

Nagar, A.L. & R.K. Das (1983): *Basic Statistics*, 2nd Edition, Oxford University Press, New Delhi.

Nath, Vikas: *Trends in Public Investment in Agriculture*, http://vikasnath.org.

National Council of Applied Economic Research (1961): *Techno-Economic Survey of Manipur*.

------ (1979): *Capital Formation in Agriculture*, December, New Delhi.

National Sample Survey Organisation (1997): *Operational Land Holdings in India 1991-92: Salient Features*, 48th Round, (January-December, 1992), Report no. 407.

------ (1998): Household Assets ad Liabilities as on 30-6-1991", *Debt and Investment Survey*, 48th Round (January-December 1992), Report no. 419.

------ (1998): "Household Capital Expenditure during 1-7-91 to 30-6-92", *Debt and Investment Survey*, 48th Round, (Jan.-Dec. 1992), Report no. 437.

Nurkse, Ragnar (1980): *Problem of Capital Formation in Underdeveloped Countries*, Oxford University Press, New Delhi.

Pal, Suresh & Alka Singh (1998): "Strengthening the Indian Agricultural Research and Extension Systems: Funding and Institutional Imperatives", *Indian Journal of Agricultural Economics*, vol. 53, no. 3, July-Sept., pp. 488-503.

Panikar, P.G.K. (1969): "Capital formation in Indian Agriculture", *Indian Journal of Agricultural Economics*, vol. 24, no. 4, Oct-Dec., pp. 31-43.

Prakash, Brahm & Sushila Srivastava (1996): "Capital Formation in Indian Agriculture: Trends, Constraints and Prospects" (Summaries), *Indian Journal of Agricultural Economics*, vol. 51, no. 4, Oct.-Dec.

Raghavan, M. (1996): "Investment Lag in Agriculture", *Economic and Political Weekly*, vol. XXX, no. 43, Oct.26, p. 2892.

Rao, C.N. Hanumantha (1998): "Agricultural Growth, Sustainability and Poverty Alleviation: Recent Trends and Major Issues of Reform", *Economic and Political Weekly*, vol. XXXIII, nos. 29 & 30, July 18, pp. 1943-1948.

Rao, J. Mohan (1998): "Food, Agriculture and Reforms Change and Continuity", *Economic and Political Weekly*, vol. XXXIII, nos. 29 & 30, July 18-24/25-31, pp. 1955-1960.

Ratha, D.K. & Atul Sarma (1992): "Price Subsidies and Irrigation Investment in India: Macro Implications", *Economic and Political Weekly*, vol. XXVII, no. 39, Sept. 26, pp. A-117-A-121.

Reserve Bank of India: "Assets and Liabilities of Rural Households as on 30th June 1971", Statistical Tables Assam, Manipur, Meghalaya and Tripura, *All India Debt and Investment Survey 1971-72*.

------ (1978a): "Capital Expenditure and Capital Formation of Rural households during 1971-72", *All India Debt and Investment Survey.*

------ (1978b): "Capital Expenditure and Capital Formation of Rural households during 1971-72", Statistical Tables Relating to Capital Expenditure and Capital Formation of Rural Households During the Year ended 30th June 1972, *All India Debt and Investment Survey*.

------ (1982): *Capital Formation and Saving in India 1950-51 to 1979-80* Report of the Working Group on Savings, Appointed by Department of Statistics, Ministry of Planning, Government of India.

------ (1987): "Statistical Tables Relating to Capital Expenditure and Capital Formation of households during the year ended 30th June 1982, *All India Debt and Investment Survey*.

Robinson, Joan (1953-54): "The Production Function and the Theory of Capital", *The Review of Economic Studies*, vol. 21(2), no. 55, pp. 81-106.

Saleth, R. Maria (1999): "Irrigation Privatisation in India Options: Issues and Experience", *Economic and Political Weekly*, vol. XXXIV, no. 26, June, 26, pp. A-86-A-92.

Sampath, R.K., B.K. Singh, & Zakir Husain (1984): "Agricultural Production Functions, Technical Progress and Relative Land Utilization by Large Farms Across Countries", *Indian Journal of Agricultural Economics*, vol. 39, no. 1, Jan.- March , pp. 53-69.

Sankhayan, P.L. (1988): *Introduction to the Economics of Agricultural Production*, Prentice-Hall of India Pvt. Ltd., New Delhi.

Sarao, B.S. (1987): *All India Report on Input Survey 1976-77*, vol. 1, Department of Agriculture and Rural Development, Ministry of Agriculture and Rural Development, New Delhi.

Sarma, J.S. (1990): "Land Utilisation" in *Handbook of Agriculture*, Indian Council of Agricultural Research, New Delhi.

Sawant, S.D. (1997): "Performance of Indian Agriculture with Special Reference to Regional Variations", *Indian Journal of Agricultural Economics*, vol. 52, no. 3, July-Sept., pp. 353-373.

Sawant, S.D., Vaidehi Peptardar & Sandhy Mhatre (2002): "Capital Formation and Growth in Agriculture Neglected Aspects and Dimensions", *Economic and*

Political Weekly, vol. XXXVII, no. 11, March, 16, pp. 1068-1072.

Schiff, Maurice & Alberto Valdes (1998): "Agriculture and the Macro economy", B. Gardner and G. Raussek (eds.): *Elsevier Science Handbook of Agricultural Economics*, World Bank, Washington D.C., March 24, Revisions.

Schultz, Theodore W. (1961): "Investment in Human Capital", *The American Economic Review*, vol. LI, March, no. 1, pp. 1-17.

------ (1964): *Transforming Traditional Agriculture*, Yale University Press, New Haven, Connecticut.

------ (1965): *Economic Crises in World Agriculture*, The University of Michigan Press.

Seetharam, G.N. (1984): *Strategy and Tactics of India's Agricultural Development: The Role of the State*, Ajanta Publications, Delhi.

Sharma, Satish (1982): *Technological Response in Developing Agriculture (Rajasthan - A case Study)*, National Publishing House, New Delhi.

Shetty, S.L. (1990): "Investment in Agriculture: Brief Review of Recent Trends", *Economic and Political Weekly*, vol. XXV, nos. 7 & 8, February 17-24, pp. 389-398.

------ (2005): Saving and Investment Estimates: Time to Take a Fresh Look, *Economic and Political Weekly*, vol. XL, no. 7, Feb. 12-18, pp. 606-610.

Shukla, Tara (1965): *Capital Formation in Indian Agriculture*, Vora and Co. Publishers Pvt. Ltd., Bombay.

Sidhu, D.S. & A.J. Singh (1986): "Technological Change in Indian Agriculture", M.L. Dantwala and Others (eds.): *Indian Agricultural Development since Independence: A Collection of Essays*, oxford & IBH Publishing Co. Pvt. Ltd., New Delhi pp. 140-161.

Singh, E. Bijoykumar (1992): *Distribution of Gains from Technological Change in Indian Agriculture: A Case Study of Manipur*, Unpublished Ph. D. Thesis, Manipur University.

Singh, E. Iboyaima (1985): *Economic Problems of Manipur*, 2nd Edition, Published by: Kunjakishwar Elangbam, Imphal.

Singh, I.J., K.N. Rai, & J.C. Karwasra (1997): "Regional Variations in Agricultural Performance in India", *Indian Journal of Agricultural Economics*, vol. 52, no. 3, July-Sept., pp. 374-386.

Singh, N. Lokeshor (1988): *Studies in Economics of Farm Management in Imphal Districts of Manipur: A Case Study of Naorem Leikai Village*, Unpublished M. Phil. Dissertation, Manipur University.

Singh, Karam (1997): "Trends and Determinants of Capital Formation in Indian Agriculture", *Indian Journal of Agricultural Economics*, vol. 52, no. 1, Jan.-March, pp. 48-51.

Singh, L. Bokul (1989): *Problems of Agricultural Labour in Manipur with Special Reference to Thoubal Districts*, Unpublished Ph. D. Thesis, Manipur University.

Singh, L. Udhob (1989): *A Brief History of Agricultural Development in Manipur (1946-1983)*, L. Saratkumar Singh & Brothers, Imphal.

Singh, N. Mohendro (1989): *Agrarian Relation in Manipur*, Unpublished Ph. D. Thesis, Manipur University.

Singh, N. Ram (1987): *Agricultural Development in the North-Eastern Region of India including Sikkim 1951-52 to 1986-87*, Unpublished Ph. D. Thesis, Manipur University.

Singh, Rachhpal, Bant Singh & Jaswinder Kaur (1996): "A Spatio-Temporal Analysis of Capital Accumulation on Punjab Farms", *Indian Journal of Agricultural Economics*, vol. 51, no. 4, Oct.-Dec., pp. 579-586.

Singh, Th. Biren (1993): *A Perspective of Agricultural and Horticultural Development in Manipur During the Plan Period*, Unpublished Ph. D. Thesis, Manipur University.

Singh, W. Kumar (1987): *Population Growth and Agricultural Development in Manipur 1951-1985*, Unpublished Ph. D. Thesis, Manipur University.

------ (1999): "Technological Change and Functional Income Distribution in Manipur Agriculture", *Indian Journal of Agricultural Economics*, vol. 54, no. 12, April-June, pp. 220-227.

Smith, V.L. (1962): "The Theory of Capital", *The American Economic Review*, vol. LII, no. 3, June, pp. 481-491.

Snodgrass, Milton M. & L.T. Wallace (1970): "Agriculture, Economics and Resource Management, Prentice-Hall, New Jersey, Published in 1964 & 1970 under the title *Agriculture, Economics & Growth*.

Srinivasamurthy, A.P.: *Investment Allocation in Indian Planning*, Himalaya Publishing House, Bombay.

Solow, R.M. (1962): "Technical Progress, Capital Formation, and Economic Growth", *The American Economic Review*, vol. LII, no. 2, May, pp. 76-86.

Sundararajan, V. & Subhash Thakur (1980): *Public Investment, Crowding out and Growth: A Dynamic Model Applied to India and Korea*, IMF Staff Papers, vol. 27, Washington, pp. 814-55.

Taneja, M.L. & R.M. Myer: *Economics of Development and Planning (with special reference to India)*, Published by Shoban-Lal Nagin Chand & Co., Jalandhar.

Thirlwall, A.P. (1980): *Growth and Development with Special Reference to Developing Economies*, 2nd Edition, The Macmillan Press Ltd., London.

Todaro, Michael, P. (1993): *Economic Development in the Third World*, First Indian Edition, Orient Longman Ltd., Hyderabad.

Upton, Martin (1976): *Agricultural Production Economics and Resource Use*, Oxford University Press, London.

Wagle, M.P. (1994): "Estimates of Aggregate Functions for Demand for Fertilisers and Private Investment in Indian Agriculture", *Indian Journal of Agricultural Economics*, vol. 49, no. 1, Jan.-March, pp. 56-69.

Index

K

L

M

N

O

P

■■■